# Praise for *Behavioral Data Analysis with R and Python*

"Whereas some data science books seek to teach their readers a list of new techniques, Florent's goal is different and more profound. He seeks to teach us data-centric wisdom: how to build a detailed and nuanced understanding of the data footprints of human behavior."

—*Steve Wendel, Head of Behavioral Science, Morningstar*

"*Behavioral Data Analysis* will help you make sense of data even when you can't run controlled experiments."

—*Colin McFarland, Dir. of Experimentation Platform, Netflix*

"We are awash with data and this is a long-needed resource to guide practitioners in using that data for building valid causal models to predict and explain real-world behaviors."

—*David Lewis, President, BEworks Research Institute at BEworks*

"For anyone looking to apply behavioral science to guide business decision making, this book provides a valuable in-depth introduction on how to leverage causal diagrams in experimentation and behavioral analysis."

—*Matt Wright, Dir. of Behavioral Science, WiderFunnel*

"Part of what makes behavioral science work is seamlessly blending quantitative and qualitative insights to support our understanding of why people do what they do. This book can help anyone with some basic data skills to meaningfully participate in that behavioral science process."

—*Matt Wallaert, Head of Behavioral Science at frog, Author of* Start at the End: How to Build Products That Create Change

# Behavioral Data Analysis with R and Python

*Customer-Driven Data for Real Business Results*

*Florent Buisson*

Beijing · Boston · Farnham · Sebastopol · Tokyo

**Behavioral Data Analysis with R and Python**

by Florent Buisson

Published by O'Reilly Media, Inc., 1005 Gravenstein Highway North, Sebastopol, CA 95472.

O'Reilly books may be purchased for educational, business, or sales promotional use. Online editions are also available for most titles (*http://oreilly.com*). For more information, contact our corporate/institutional sales department: 800-998-9938 or *corporate@oreilly.com*.

| | |
|---|---|
| **Acquisitions Editor:** Michelle Smith | **Indexer:** Sue Klefstad |
| **Development Editor:** Gary O'Brien | **Interior Designer:** David Futato |
| **Production Editor:** Deborah Baker | **Cover Designer:** Karen Montgomery |
| **Copyeditor:** Charles Roumeliotis | **Illustrator:** Kate Dullea |
| **Proofreader:** Kim Sandoval | |

June 2021: First Edition

**Revision History for the First Edition**
2021-06-15: First Release
2023-05-12: Second Release
2023-09-15: Third Release

See *http://oreilly.com/catalog/errata.csp?isbn=9781492061373* for release details.

978-1-492-06137-3

[LSI]

# Table of Contents

Preface.................................................................. xi

## Part I.    Understanding Behaviors

1. The Causal-Behavioral Framework for Data Analysis............................ 3
   Why We Need Causal Analytics to Explain Human Behavior          4
      The Different Types of Analytics                            4
      Human Beings Are Complicated                                5
   Confound It! The Hidden Dangers of Letting Regression Sort It Out      8
      Data                                                        9
      Why Correlation Is Not Causation: A Confounder in Action     9
      Too Many Variables Can Spoil the Broth                      11
   Conclusion                                                     17

2. Understanding Behavioral Data............................................ 19
   A Basic Model of Human Behavior                                20
      Personal Characteristics                                    21
      Cognition and Emotions                                      23
      Intentions                                                  24
      Actions                                                     25
      Business Behaviors                                          26
   How to Connect Behaviors and Data                              28
      Develop a Behavioral Integrity Mindset                      28
      Distrust and Verify                                         29
      Identify the Category                                       30
      Refine Behavioral Variables                                 32
      Understand the Context                                      33

Conclusion                                                                                          35

## Part II.    Causal Diagrams and Deconfounding

### 3. Introduction to Causal Diagrams. . . . . . . . . . . . . . . . . . . . . . . . . . . . . . . . . . . . . . . . . . . .  39
Causal Diagrams and the Causal-Behavioral Framework              40
    Causal Diagrams Represent Behaviors                                        41
    Causal Diagrams Represent Data                                              42
Fundamental Structures of Causal Diagrams                               46
    Chains                                                                                    46
    Forks                                                                                      50
    Colliders                                                                                 52
Common Transformations of Causal Diagrams                           53
    Slicing/Disaggregating Variables                                          54
    Aggregating Variables                                                           55
    What About Cycles?                                                             57
    Paths                                                                                      60
Conclusion                                                                                          61

### 4. Building Causal Diagrams from Scratch. . . . . . . . . . . . . . . . . . . . . . . . . . . . . . . . . .  63
Business Problem and Data Setup                                              64
    Data and Packages                                                               64
    Understanding the Relationship of Interest                          65
Identify Candidate Variables to Include                                    67
    Actions                                                                                   69
    Intentions                                                                               70
    Cognition and Emotions                                                       71
    Personal Characteristics                                                       72
    Business Behaviors                                                               75
    Time Trends                                                                           75
Validate Observable Variables to Include Based on Data           77
    Relationships Between Numeric Variables                            78
    Relationships Between Categorical Variables                       81
    Relationships Between Numeric and Categorical Variables   84
Expand Causal Diagram Iteratively                                          86
    Identify Proxies for Unobserved Variables                          86
    Identify Further Causes                                                        87
    Iterate                                                                                     88
Simplify Causal Diagram                                                           88
Conclusion                                                                                          90

5.  Using Causal Diagrams to Deconfound Data Analyses. . . . . . . . . . . . . . . . . . . . . . . . . **91**

  Business Problem: Ice Cream and Bottled Water Sales          92
  The Disjunctive Cause Criterion                              94
      Definition                                               94
      First Block                                              95
      Second Block                                             97
  The Backdoor Criterion                                       97
      Definitions                                              98
      First Block                                              100
      Second Block                                             101
  Conclusion                                                   103

## Part III.   Robust Data Analysis

6.  Handling Missing Data. . . . . . . . . . . . . . . . . . . . . . . . . . . . . . . . . . . . . . . . . . **107**

  Data and Packages                                            109
  Visualizing Missing Data                                     110
      Amount of Missing Data                                   113
      Correlation of Missingness                               115
  Diagnosing Missing Data                                      121
      Causes of Missingness: Rubin's Classification            124
      Diagnosing MCAR Variables                                126
      Diagnosing MAR Variables                                 128
      Diagnosing MNAR Variables                                130
      Missingness as a Spectrum                                132
  Handling Missing Data                                        136
      Introduction to Multiple Imputation (MI)                 137
      Default Imputation Method: Predictive Mean Matching      140
      From PMM to Normal Imputation (R Only)                   141
      Adding Auxiliary Variables                               143
      Scaling Up the Number of Imputed Data Sets               145
  Conclusion                                                   146

7.  Measuring Uncertainty with the Bootstrap. . . . . . . . . . . . . . . . . . . . . . . . . . . . **147**

  Intro to the Bootstrap: "Polling" Oneself Up                148
      Packages                                                 148
      The Business Problem: Small Data with an Outlier         148
      Bootstrap Confidence Interval for the Sample Mean        150
      Bootstrap Confidence Intervals for Ad Hoc Statistics     155
  The Bootstrap for Regression Analysis                        157
  When to Use the Bootstrap                                    160

Conditions for the Traditional Central Estimate to Be Sufficient          161
Conditions for the Traditional CI to Be Sufficient          162
Determining the Number of Bootstrap Samples          164
Optimizing the Bootstrap in R and Python          166
R: The BehavioralDataAnalysis Package          166
Python Optimization          169
Conclusion          170

## Part IV.   Designing and Analyzing Experiments

**8. Experimental Design: The Basics.** . . . . . . . . . . . . . . . . . . . . . . . . . . . . . . . . . . . . **173**
Planning the Experiment: Theory of Change          174
Business Goal and Target Metric          175
Intervention          177
Behavioral Logic          179
Data and Packages          181
Determining Random Assignment and Sample Size/Power          182
Random Assignment          182
Sample Size and Power Analysis          185
Analyzing and Interpreting Experimental Results          199
Conclusion          202

**9. Stratified Randomization.** . . . . . . . . . . . . . . . . . . . . . . . . . . . . . . . . . . . . . . . . **203**
Planning the Experiment          205
Business Goal and Target Metric          205
Definition of the Intervention          207
Behavioral Logic          208
Data and Packages          208
Determining Random Assignment and Sample Size/Power          209
Random Assignment          209
Power Analysis with Bootstrap Simulations          216
Analyzing and Interpreting Experimental Results          223
Intention-to-Treat Estimate for Encouragement Intervention          224
Complier Average Causal Estimate for Mandatory Intervention          226
Conclusion          231

**10. Cluster Randomization and Hierarchical Modeling.** . . . . . . . . . . . . . . . . . . . . . . . **233**
Planning the Experiment          234
Business Goal and Target Metric          234
Definition of the Intervention          234
Behavioral Logic          236

Data and Packages                                                     236
Introduction to Hierarchical Modeling                                 237
  R Code                                                     238
  Python Code                                                240
Determining Random Assignment and Sample Size/Power                   242
  Random Assignment                                          242
  Power Analysis                                             245
Analyzing the Experiment                                              253
Conclusion                                                            253

## Part V.   Advanced Tools in Behavioral Data Analysis

**11. Introduction to Moderation** . . . . . . . . . . . . . . . . . . . . . . . . . . . . . . . . . **257**
Data and Packages                                                     258
Behavioral Varieties of Moderation                                    258
  Segmentation                                               259
  Interactions                                               265
  Nonlinearities                                             266
How to Apply Moderation                                               269
  When to Look for Moderation?                               270
  Multiple Moderators                                        281
  Validating Moderation with Bootstrap                       287
  Interpreting Individual Coefficients                       289
Conclusion                                                            295

**12. Mediation and Instrumental Variables** . . . . . . . . . . . . . . . . . . . . . . . . . . **297**
Mediation                                                             298
  Understanding Causal Mechanisms                            298
  Causal Biases                                              299
  Identifying Mediation                                      301
  Measuring Mediation                                        302
Instrumental Variables                                                307
  Data                                                       307
  Packages                                                   308
  Understanding and Applying IVs                             308
  Measurement                                                311
  Applying IVs: Frequently Asked Questions                   314
Conclusion                                                            315

Bibliography . . . . . . . . . . . . . . . . . . . . . . . . . . . . . . . . . . . . . . . . . . . . . . . . . . . . . . . . . . . . 317

Index . . . . . . . . . . . . . . . . . . . . . . . . . . . . . . . . . . . . . . . . . . . . . . . . . . . . . . . . . . . . . . . . . . 321

# Preface

*Statistics is a subject of amazingly many uses and surprisingly few effective practitioners.*
—Bradley Efron and R. J. Tibshirani, *An Introduction to the Bootstrap* (1993)

Welcome to *Behavioral Data Analysis with R and Python*! That we live in the age of data has become a platitude. Engineers now routinely use data from sensors on machines and turbines to predict when these will fail and do preventive maintenance. Similarly, marketers use troves of data, from your demographic information to your past purchases, to determine which ad to serve you and when. As the phrase goes, "Data is the new oil," and algorithms are the new combustion engine powering our economy forward.

Most books on analytics, machine learning, and data science implicitly presume that the problems that engineers and marketers are trying to solve can be handled with the same approaches and tools. Sure, the variables have different names and there is some domain-specific knowledge to acquire, but k-means clustering is k-means clustering, whether you're clustering data about turbines or posts on social media. By adopting machine learning tools wholesale this way, businesses have often been able to accurately predict behaviors, but at the expense of a deeper and richer understanding of what's actually going on. This has fed into the criticism of data science models as "black boxes."

Instead of aiming for accurate but opaque predictions, this book strives to answer the question, "What drives behavior?" If we decide to send an email to prospective customers, will they buy a subscription to our service *as a result of the email*? And which groups of customers should receive the email? Do older customers tend to purchase different products *because* they're older? What is the impact of customer experience on loyalty and retention? By shifting our perspective from predicting behaviors to explaining them and measuring their causes, we'll be able to break the curse of "correlation is not causation," which has prevented generations of analysts from being confident in the results of their models.

This shift won't come from the introduction of new analytical tools: we'll use only two data analysis tools: good old linear regression and its logistic sibling. These two are intrinsically more readable than other types of models. Certainly, this often comes at the cost of a lower predictive accuracy (i.e., they make more and larger errors in prediction), but that doesn't matter for our purpose here of measuring relationships between variables.

Instead we'll spend a lot of time learning to make sense of data. In my role as a data science interviewer, I have seen many candidates who can use sophisticated machine learning algorithms but haven't developed a strong sense for data: they have little intuition for what's going on in their data apart from what their algorithms tell them.

I believe that you can develop that intuition, and along the way increase the value and outcomes of your analytics projects—often dramatically—by adopting the following:

- A behavioral science mindset, which sees data not as an end in itself but as a lens into the psychology and behaviors of human beings
- A causal analytics toolkit, which allows us to confidently say that one thing causes another and to determine how strong that relationship is

While each of these can provide great benefits on its own, I believe that they are natural complements that are best used together. Given that "a behavioral science mindset using a causal analytics toolkit" is a bit of a mouthful, I'll call it instead a causal-behavioral approach or framework. This framework has an added benefit: it applies equally to experimental and historical data while leveraging the differences between them. This contrasts with traditional analytics, which handles them with completely different tools (e.g., ANOVA and T-test for experimental data) and data science, which doesn't treat experimental data differently from historical data.

## Who This Book Is For

If you're analyzing data in a business with R or Python, this book is for you. I use the word "business" loosely to mean any for-profit, nonprofit, or governmental organization where correct insights and actionable conclusions driving action are what matters.

In terms of math and stats background, it doesn't matter whether you are a business analyst building monthly forecasts, a UX researcher looking at click-through behaviors, or a data scientist building machine learning models. This book has one fundamental prerequisite: you need to be at least somewhat familiar with linear and logistic regression. If you understand regression, you can follow the argument of this book and reap great benefits from it. On the other side of the spectrum, I believe even expert data scientists with PhDs in statistics or computer science will find the material new and useful, provided they are not already specialists in behavioral or causal analytics.

In terms of programming background, you need to be able to read and write code in R or Python, ideally both. I will not show you how to define a function or how to manipulate data structures such as data frames or pandas. There are already excellent books doing a better job of it than I would (e.g., *Python for Data Analysis* by Wes McKinney (O'Reilly) and *R for Data Science* by Garrett Grolemund and Hadley Wickham (O'Reilly)). If you've read any of these books, taken an introductory class, or used at least one of the two languages at work, then you'll be equipped for the material here. Similarly, I will usually not present and discuss the code used to create the numerous figures in the book, although it will be in the book's GitHub.

## Who This Book Is Not For

If you're in academia or a field that requires you to follow academic norms (e.g., pharmaceutical trials), this book might still be of interest to you—but the recipes I'm describing might get you in trouble with your advisor/editor/manager.

This book is *not* an overview of conventional behavioral data analysis methods, such as T-test or ANOVA. I have yet to encounter a situation where regression was less effective than these methods for providing an answer to a business question, which is why I'm deliberately restraining this book to linear and logistic regression. If you want to learn other methods, you'll have to look elsewhere (e.g., *Hands-On Machine Learning with Scikit-Learn, Keras, and TensorFlow* (O'Reilly) by Aurélien Géron for machine learning algorithms).

Understanding and changing behaviors in applied settings requires both data analysis and qualitative skills. This book focuses squarely on the former, primarily for reasons of space. In addition, there are already excellent books that cover the latter, such as *Nudge: Improving Decisions About Health, Wealth, and Happiness* (Penguin) by Richard Thaler and Cass Sunstein and *Designing for Behavior Change: Applying Psychology and Behavioral Economics* (O'Reilly) by Stephen Wendel. Nonetheless, I'll provide an introduction to behavioral science concepts so that you can apply the tools from this book even if you're new to the field.

Finally, if you're completely new to data analysis in R or Python, this is not the book for you. I recommend starting with some of the excellent introductions out there, such as the ones mentioned in this section.

# R and Python Code

Why R *and* Python? Why not whichever of the two is superior? The "R versus Python" debate is still alive and kicking on the Internet. It is also, in my humble opinion, mostly irrelevant. The reality is that you'll have to use whatever language is used in your organization, period. I once worked in a healthcare company where, for historical and regulatory reasons, SAS was the dominant language. I regularly used R and Python for my own analyses, but since I couldn't avoid dealing with the legacy SAS code, I taught myself as much SAS as I needed during my first month there. Unless you spend your entire career in a company that doesn't use R or Python, you'll most likely end up picking up some basics in both anyway, so you might as well embrace bilingualism. I have yet to encounter anyone who stated that "learning to read code in [the other language] was a waste of my time."

Assuming that you have the good luck of being in an organization that uses both, which language should you work with? I think it really depends on your context and the tasks you have to do. For example, I personally prefer doing exploratory data analysis (EDA) in R, but I find that Python is much easier to use for webscraping. I advise choosing based on the specifics of your job and relying on up-to-date information: both languages are constantly improving, and what was true for a previous version of R or Python may not be true for the current version. For example, Python is becoming a much friendlier environment for EDA than it once was. Your energy is better spent on learning both than on scouring forums to pick the best of the two.

## Code Environments

I have created an R package for this book, which includes some functions not otherwise available. This package is available on GitHub (*https://github.com/BehavioralDataAnalysis/R_package*), and you can install it directly from RStudio:

```
## R
# skip the first line if you know that devtools is already installed on
# your computer. It won't hurt to run it if you're unsure.
install.packages("devtools")
devtools::install_github("BehavioralDataAnalysis/R_package")
```

At the beginning of each chapter, I'll call out the R and Python packages that need to be loaded specifically for that chapter. In addition, I'll also be using a few standard packages across the entire book; to avoid repetition, these are called out only here (they are already included in all the scripts on GitHub). You should always start your code with them as well as with a few parameter settings:

```
## R
library(tidyverse)
library(boot) #Required for Bootstrap simulations
library(rstudioapi) #To load data from local folder
library(ggpubr) #To generate multi-plots
```

```
# Setting the random seed will ensure reproducibility of random numbers
set.seed(1234)
# I personally find the default scientific number notation (i.e. with
# exponents) less readable in results, so I cancel it
options(scipen=10)

## Python
import pandas as pd
import numpy as np
import statsmodels.formula.api as smf
from statsmodels.formula.api import ols
import matplotlib.pyplot as plt # For graphics
import seaborn as sns # For graphics
```

## Code Conventions

I use R in RStudio. R 4.0 was launched while I was writing this book and I have adopted it to keep the book as current as possible.

R code is written in a code font with a comment indicating the language used, like this:

```
## R
> x <- 3
> x
[1] 3
```

I use Python in Anaconda's Spyder. The "Python 2.0 vs. 3.0" discussion is hopefully behind us (at least for new code; legacy code can be a different story), and I'll be using Python 3.7. The convention for Python code is somewhat similar to the one for R:

```
## Python
In [1]: x = 3
In [2]: x
Out[2]: 3
```

We'll often look at the output of regressions. It can be quite verbose, with a lot of diagnostics that are not relevant to the arguments of this book. You shouldn't disregard them in real life, but that's a matter better covered in other books. Therefore, I'll abbreviate the output like this:

```
## R
> model1 <- lm(icecream_sales ~ temps, data=stand_dat)
> summary(model1)

...
Coefficients:
            Estimate Std. Error t value Pr(>|t|)
(Intercept) -4519.055    454.566  -9.941   <2e-16 ***
temps        1145.320      7.826 146.348   <2e-16 ***
...
```

```
## Python
model1 = ols("icecream_sales ~ temps", data=stand_data_df)
print(model1.fit().summary())

...
                coef    std err         t     P>|t|      [0.025     0.975]
--------------------------------------------------------------------------
Intercept   -4519.0554   454.566    -9.941    0.000    -5410.439  -3627.672
temps        1145.3197     7.826   146.348    0.000     1129.973   1160.666
...
```

# Functional-Style Programming 101

One of the steps of going from beginner to intermediate level as a programmer is to stop writing scripts in which your code is just a long succession of instructions and to structure your code into functions instead. In this book, we'll write and reuse functions across chapters, such as the following ones to build Bootstrap confidence intervals:

```
## R
boot_CI_fun <- function(dat, metric_fun, B=20, conf.level=0.9){

  boot_vec <- sapply(1:B, function(x){
    cat("bootstrap iteration ", x, "\n")
    metric_fun(slice_sample(dat, n = nrow(dat), replace = TRUE))})
  boot_vec <- sort(boot_vec, decreasing = FALSE)
  offset = round(B * (1 - conf.level) / 2)
  CI <- c(boot_vec[offset], boot_vec[B+1-offset])
  return(CI)
}

## Python
def boot_CI_fun(dat_df, metric_fun, B = 20, conf_level = 9/10):

  coeff_boot = []

  # Calculate coeff of interest for each simulation
  for b in range(B):
      print("beginning iteration number " + str(b) + "\n")
      boot_df = dat_df.groupby("rep_ID").sample(n=1200, replace=True)
      coeff = metric_fun(boot_df)
      coeff_boot.append(coeff)

  # Extract confidence interval
  coeff_boot.sort()
  offset = round(B * (1 - conf_level) / 2)
  CI = [coeff_boot[offset], coeff_boot[-(offset+1)]]

  return CI
```

Functions also have the added advantage of limiting incomprehension spillovers: even if you don't understand how the preceding functions work, you can still take for granted that they return confidence intervals and follow the rest of the reasoning, postponing a deeper dive into their code until later.

## Using Code Examples

Supplemental material (code examples, etc.) is available for download at *https://oreil.ly/BehavioralDataAnalysis*.

If you have a technical question or a problem using the code examples, please send email to *support@oreilly.com*.

This book is here to help you get your job done. In general, if example code is offered with this book, you may use it in your programs and documentation. You do not need to contact us for permission unless you're reproducing a significant portion of the code. For example, writing a program that uses several chunks of code from this book does not require permission. Selling or distributing examples from O'Reilly books does require permission. Answering a question by citing this book and quoting example code does not require permission. Incorporating a significant amount of example code from this book into your product's documentation does require permission.

We appreciate, but do not require, attribution. An attribution usually includes the title, author, publisher, and ISBN. For example: "*Behavioral Data Analysis with R and Python*, by Florent Buisson (O'Reilly). Copyright 2021 Florent Buisson, 978-1-492-06137-3."

If you feel your use of code examples falls outside fair use or the permission given above, feel free to contact us at *permissions@oreilly.com*.

## Navigating This Book

The core intuition of the book is the idea that effective data analysis relies on a constant back and forth between three things:

- Actual behaviors in the real world and related psychological phenomena, such as intentions, thoughts, and emotions
- Causal analytics and especially causal diagrams
- Data

The book is split into five parts:

*Part I, "Understanding Behaviors"*
This part sets the stage with the causal-behavioral framework and the connections between behaviors, causal reasoning, and data.

*Part II, "Causal Diagrams and Deconfounding"*
This part introduces the concept of confounding and explains how causal diagrams allow us to deconfound our data analyses.

*Part III, "Robust Data Analysis"*
Here we explore tools for missing data and introduce Bootstrap simulations, as we'll rely extensively on Bootstrap confidence intervals in the rest of the book. Data that is small, incomplete, or irregularly shaped (e.g., with multiple peaks or outliers) is not a new problem, but it can be especially acute with behavioral data.

*Part IV, "Designing and Analyzing Experiments"*
In this part, we'll discuss how to design and analyze experiments.

*Part V, "Advanced Tools in Behavioral Data Analysis"*
Finally, we bring everything together to explore moderation, mediation, and instrumental variables.

The various parts of the book build somewhat on each other, and consequently I recommend reading them in order, at least for your first pass.

## Conventions Used in This Book

The following typographical conventions are used in this book:

*Italic*
Indicates new terms, URLs, email addresses, filenames, and file extensions.

`Constant width`
Used for program listings, as well as within paragraphs to refer to program elements such as variable or function names, databases, data types, environment variables, statements, and keywords.

**`Constant width bold`**
Shows commands or other text that should be typed literally by the user.

*`Constant width italic`*
Shows text that should be replaced with user-supplied values or by values determined by context.

 This element signifies a tip or suggestion.

This element signifies a general note.

 This element indicates a warning or caution.

## O'Reilly Online Learning

 For more than 40 years, *O'Reilly Media* has provided technology and business training, knowledge, and insight to help companies succeed.

Our unique network of experts and innovators share their knowledge and expertise through books, articles, and our online learning platform. O'Reilly's online learning platform gives you on-demand access to live training courses, in-depth learning paths, interactive coding environments, and a vast collection of text and video from O'Reilly and 200+ other publishers. For more information, visit *http://oreilly.com*.

## How to Contact Us

Please address comments and questions concerning this book to the publisher:

O'Reilly Media, Inc.
1005 Gravenstein Highway North
Sebastopol, CA 95472
800-889-8969 (in the United States or Canada)
707-829-7019 (international or local)
707-829-0104 (fax)
*support@oreilly.com*
*https://www.oreilly.com/about/contact.html*

You can access the web page for this book, which lists errata, examples, and additional information at *https://oreil.ly/Behavioral_Data_Analysis_with_R_and_Python*.

For news and information about our books and courses, visit *http://oreilly.com*.

Find us on LinkedIn: *https://linkedin.com/company/oreilly-media*

Follow us on Twitter: *https://twitter.com/oreillymedia*

Watch us on YouTube: *https://youtube.com/oreillymedia*

## Acknowledgments

Authors often thank their spouses for their patience and call out especially insightful reviewers. I have been blessed with having both in the same person. I don't think anyone else would have dared or managed to send me back to the drawing board so many times, and this book is tremendously better for it. So my first thanks goes to my life and thought partner.

Several of my colleagues and fellow behavioral scientists have been generous enough to devote their time to reading and commenting on an earlier draft. This book is all the better for it. Thank you (in reverse alphabetical order) Jean Utke, Jessica Jakubowski, Chinmaya Gupta, and Phaedra Daipha!

A special thank goes to Bethany Winkel for her help with the writing.

I now cringe at how rough and confusing the very first drafts were. My development editor and technical reviewers patiently prodded me all the way to where the book is now, sharing their wealth of perspective and expertise. Thank you Gary O'Brien, and thank you Xuan Yin, Shannon White, Jason Stanley, Matt LeMay, and Andreas Kaltenbrunner.

# Understanding Behaviors

In this first part of the book, I'll explain why behavioral data analysis requires a new approach.

Chapter 1 will describe that new approach, the causal-behavioral framework for data analysis. We'll look at a concrete example showing how even the simplest of data analyses can be derailed by the presence of a confounder. Solving that issue is at best complicated and at worst impossible with traditional approaches, but the new framework provides a straightforward process.

Chapter 2 will explore further the specificities of behavioral data while providing both a gentle introduction to behavioral science and a process for ensuring that our data adequately reflects the corresponding real-life behaviors.

# CHAPTER 1

# The Causal-Behavioral Framework for Data Analysis

As we discussed in the preface, understanding what drives behaviors in order to change them is one of the key goals of applied analytics, whether in a business, a nonprofit, or a public organization. We want to figure out why someone bought something and why someone else *didn't* buy it. We want to understand why someone renewed their subscription, contacted a call center instead of paying online, registered to be an organ donor, or gave to a nonprofit. Having this knowledge allows us to predict what people will do under different scenarios and helps us to determine what our organization can do to encourage them to do it again (or not). I believe that this goal is best achieved by combining data analysis with a behavioral science mindset and a causal analytics toolkit to create an integrated approach I have dubbed the "causal-behavioral framework." In this framework, *behaviors* are at the top because understanding them is our ultimate goal. This understanding is achieved by using *causal diagrams* and *data*, which form the two supporting pillars of the triangle (Figure 1-1).

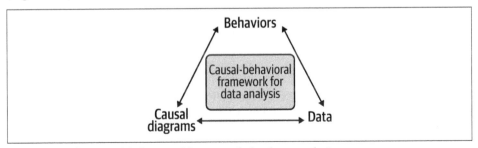

*Figure 1-1. The causal-behavioral framework for data analysis*

Over the course of the book, we'll explore each leg of the triangle and see how they connect to each other. In the final chapter, we'll see all of our work come together by achieving with one line of code what would be a daunting task with traditional approaches: measuring the degree to which customer satisfaction increases future customer spending. In addition to performing such extraordinary feats, this new framework will also allow you to more effectively perform common analyses such as determining the effect of an email campaign or a product feature on purchasing behavior.

Before getting to that, readers familiar with predictive analytics may wonder why I'm advocating for causal analytics instead. The answer is that even though predictive analytics have been (and will remain) very successful in business settings, they can fall short when your analyses pertain to human behaviors. In particular, adopting a causal approach can help us identify and resolve "confounding," a very common problem with behavioral data. I'll elaborate on these points in the rest of this first chapter.

# Why We Need Causal Analytics to Explain Human Behavior

Understanding where causal analytics fits into the analytics landscape will help us better identify why it is needed in business settings. As we'll see, that need stems from the complexity of human behavior.

## The Different Types of Analytics

There are three different types of analytics: descriptive, predictive, and causal. Descriptive analytics provides a *description* of data. In simple terms, I think of it as "what is" or "what we've measured" analytics. Business reporting falls under that umbrella. How many customers canceled their subscriptions last month? How much profit did we make last year? Whenever we're calculating an average or other simple metrics, we're implicitly using descriptive analytics. Descriptive analytics is the simplest form of analytics, but it is also underappreciated. Many organizations actually struggle to get a clear and unified view of their operations. To see the extent of that problem in an organization, just ask the same question of the finance department and the operations department and measure how different the answers are.[1]

Predictive analytics provides a *prediction*. I think of it as "what will be, assuming current conditions persist" or "what we haven't yet measured" analytics. Most machine

---

[1] To be fair, in many circumstances, they *should* be different, because the data is used for different purposes and obeys different conventions. But even questions that you would expect to have a single true answer (e.g., "How many employees do we have right now?") will generally show discrepancies.

---

learning methods (e.g., neural networks and gradient boosting models) belong to this type of analytics and help us answer questions like "How many customers will cancel their subscription next month?" and "Is that order fraudulent?" Over the past few decades, predictive analytics has transformed the world; the legions of data scientists employed in business are a testament to its success.

Finally, causal analytics provides the *causes* of data. I think of it as "what if?" or "what will be, under different conditions" analytics. It answers questions such as "How many customers will cancel their subscription next month *unless we send them a coupon*?" The most well-known tool of causal analytics is the A/B test, a.k.a. a randomized experiment or randomized controlled trial (RCT). That's because the simplest and most effective way to answer the preceding question is to send a coupon to a randomly selected group of customers and see how many of them cancel their subscription compared to a control group.

We'll cover experimentation in Part IV of the book, but before that, in Part II, we'll look at another tool from that toolkit, namely causal diagrams, which can be used even when we can't experiment. Indeed, it is one of my goals to get you to think more broadly about causal analytics rather than just equate it with experimentation.

 While these labels may give the impression of a neat categorization, in reality, there is more of a gradient between these three categories, and questions and methods get blurred between them. You may also encounter other terms, such as *prescriptive analytics*, that further blur the lines and add other nuances without dramatically altering the overall picture.

## Human Beings Are Complicated

If predictive analytics has been so successful and causal analytics uses the same data analysis tools like regression, why not stick with predictive analytics? In short, because human beings are more complicated than wind turbines. Human behavior:

*Has multiple causes*
A turbine's behavior is not influenced by its personality, the social norms of the turbine community, or the circumstances of its upbringing, whereas the predictive power of any single variable on human behavior is almost always disappointing because of those factors.

*Is context-dependent*
Minor or cosmetic alterations to the environment, such as a change in the default option of a choice, can have large impacts on behavior. This is a blessing from a behavioral *design* perspective because it allows us to drive changes in behaviors, but it's a curse from a behavioral *analytics* perspective because it means that every situation is unique in ways that are hard to predict.

*Is variable (scientists would say nondeterministic)*

The same person may behave very differently when placed repeatedly in what seems like exactly the same situation, even after controlling for cosmetic factors. This may be due to transient effects, such as moods, or long-term effects, such as getting bored with having the same breakfast every day. Both of these can dramatically change behavior but are hard to capture.

*Is innovative*

When conditions in the environment change, a human can switch to a behavior they have literally never exhibited before, and it happens even under the most mundane circumstances. For example, there's a car accident ahead on your normal commuting path and so you decide at the last minute to take a right turn.

*Is strategic*

Humans infer and react to the behaviors and intentions of others. In some cases, that can mean "repairing" a cooperation that was derailed by external circumstances, making it more robustly predictable. But in other cases, it can involve voluntarily obfuscating one's behavior to make it unpredictable when playing a competitive game like chess (or fraud!).

All these aspects of human behavior make it less predictable than that of physical objects. To find regularities that are more reliable for analysis, we must go one level deeper to understand and measure the causes of behavior. The fact that someone had oatmeal for breakfast and took a certain route on Monday doesn't mean that they will do the same on Tuesday, but you can be more confident that they'll have *some* breakfast and will take *some* route to their work.

---

## Extrapolation in Analytics, the Curse of Dimensionality, and the Lucas Critique

Readers with a quantitative background may not be fully satisfied with my statement that "human behavior is hard to predict because it's complicated," so here's the mathier version of the argument. I'll start by describing the difference between interpolation and extrapolation. Figure 1-2 shows some simulated data with a linear relationship between two variables.

The line in the figure is the regression line of best fit, the line corresponding to the linear regression between the two variables, with a slope of approximately 3. We can use it to predict unknown Y values based on a known X value (and vice versa). For example, given the value X = 50, we would predict that Y is equal to 150. There are observed points to the left of that value, that is, points for which X < 50, as well as points to the right of that value, for which X > 50. This predictive process is called interpolation because our point is between observed points (the prefix *inter* means "between"; e.g., *international* = "between nations"). Conversely, if we used the regression line with X = 0 to predict that Y = 0, this would be called extrapolation, as the

---

point we're trying to predict is outside of the cloud of observed points (*extra* means "outside"; e.g., *extraordinary* = "outside of the ordinary"). In statistics and in everyday life, to extrapolate is to leave the realm of the observed and the known in order to make predictions. Whereas interpolation is usually safe and reliable, extrapolation is always somewhat speculative: it takes a leap of faith to assume that the rules that applied within certain boundaries will still apply outside of them.

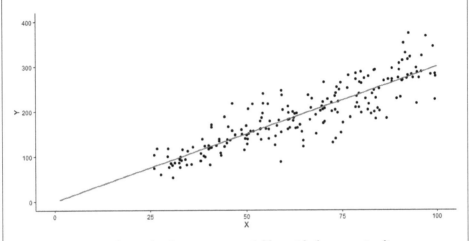

*Figure 1-2. Linear relationship between two variables, with the regression line*

Physical objects like a wind turbine are affected by a reasonably small and constant number of factors (it's not like some laws of physics take days off or new ones appear randomly). Therefore we have a lot of data points relative to the dimensions of the problem space, which means that we're almost always interpolating. For simplicity's sake, a model may neglect secondary or rare phenomena, such as a 1-in-100-year storm, but even when such outliers occur, the outcome remains somewhat predictable: the blade will break and fall in the water, not fly away.

On the other hand, human behavior is affected by a large number of different factors, which may or may not be relevant on a given day and may grow or fade over time. Therefore we usually end up having few data points relative to the dimensions of the problem space, which means that we're much more frequently extrapolating—an issue known in statistics as the "curse of dimensionality." In addition, minor changes in the environment can lead to major changes in behavior, which makes trying to predict future human behavior based on past behavior alone a gamble with poor odds.

For people interested in the genealogy of behavioral economics, the macroeconomist Robert Lucas articulated that point in the 1970s (the "Lucas critique" of Keynesian models). He recommended instead identifying the deeper parameters of human behaviors such as consumer preferences, another version of the argument I made earlier.

# Confound It! The Hidden Dangers of Letting Regression Sort It Out

I mentioned in the previous section that causal analytics often uses the same tools as predictive analytics. However, because they have different goals, the tools are used in different ways. Since regression is one of the main tools for both types of analytics, it can be a great way to illustrate the difference between predictive and causal analytics. A regression appropriate for predictive analytics would often make a terrible regression for causal analytics purposes, and vice versa.

A regression for predictive analytics is used to estimate an unknown value (often, but not always, in the future). It does this by taking known information and using a variety of factors to triangulate the best guess value for a given variable. What is important is the predicted value and its accuracy, not why or how it was predicted.

Causal analytics also uses regression, but the focus is not on estimating a value of the target variable. Instead, the focus is on the cause of that value. In regression terms, our interest is no longer in the dependent variable itself but in its relationship with a given independent variable. With a correctly structured regression, the coefficient of correlation can be a portable measure of the causal effect of an independent variable on a dependent variable.

But what does it mean to have a correctly structured regression for that purpose? Why can't we just take the regressions we already use for predictive analytics and treat the provided coefficients as measures of the causal relationship? We can't do that because each variable in the regression has the potential to modify the coefficients of other variables. Therefore our variable mix has to be crafted not to create the most accurate prediction but to create the most accurate coefficients. The two sets of variables are generally different because a variable can be highly correlated with our target variable (and therefore be highly predictive) without actually affecting that variable.

In this section, we will explore why this difference in perspective matters and why variable selection is more than half the battle in behavioral analytics. We'll do so with a concrete example from C-Mart, a fictional supermarket chain with stores across the United States. The first of two fictional companies used throughout the book, C-Mart will help us understand the opportunities and challenges of data analysis for brick-and-mortar companies in the digital age.

# Data

The GitHub folder for this chapter (*https://oreil.ly/BehavioralDataAnalysisCh1*) contains two CSV files, *chap1-stand_data.csv* and *chap1-survey_data.csv* with the data sets for the two examples in this chapter.

Table 1-1 shows the information contained by the CSV file *chap1-stand_data.csv* at the daily level about sales of ice cream and iced coffee in C-Mart's stands.

*Table 1-1. Sales information in chap1-stand_data.csv*

| Variable name | Variable description |
|---|---|
| *IceCreamSales* | Daily sales of ice cream in C-Mart's stands |
| *IcedCoffeeSales* | Daily sales of iced coffee in C-Mart's stands |
| *SummerMonth* | Binary variable indicating whether the day is in the summer months |
| *Temp* | The average temperature for that day and that stand |

Table 1-2 shows the information contained in the CSV file *chap1-survey_data.csv* from a survey of passersby outside of C-Mart's stands.

*Table 1-2. Survey information in chap1-survey_data.csv*

| Variable name | Variable description |
|---|---|
| *VanillaTaste* | Interviewee's taste for vanilla, 0-25 |
| *ChocTaste* | Interviewee's taste for chocolate, 0-25 |
| *Shopped* | Binary variable indicating whether the interviewee has ever shopped at the local C-Mart stand |

## Why Correlation Is Not Causation: A Confounder in Action

C-Mart has an ice cream stand in each store. It is the company's belief that the weather influences daily sales—or, to cast it in causality jargon, that the weather is a cause of sales. In other words, everything else being equal, we assume that people are more likely to buy ice cream on hotter days, which makes intuitive sense. This belief is supported by a strong correlation in historical data between temperature and sales as shown in Figure 1-3 (the corresponding data and code are on the book's GitHub).

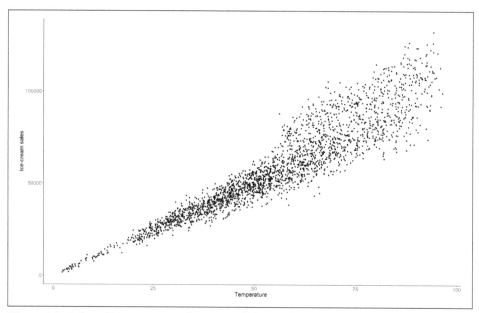

*Figure 1-3. Sales of ice cream as a function of observed temperature*

As indicated in the Preface, we'll be using regression as our main tool for data analysis. Running a linear regression of the sales of ice cream on the temperature takes a single line of code:

```
## Python (output not shown)
print(ols("icecream_sales ~ temps", data=stand_data_df).fit().summary())

## R
> summary(lm(icecream_sales ~ temps, data=stand_dat))
...
Coefficients:
            Estimate Std. Error t value Pr(>|t|)
(Intercept) -6169.844   531.506  -11.61   <2e-16 ***
temps        1171.335     9.027  129.76   <2e-16 ***
...
```

For our purposes in this book, the most relevant piece of the output is the *coefficients* section, which tells us that the estimated intercept—the theoretical average ice cream sales for a temperature of zero—is −6,169.8, which is obviously a nonsensical extrapolation. It also tells us that the estimated coefficient for the temperature is 1,171, which means that each additional degree of temperature is expected to increase sales of ice cream by $1,171.

Now, let's imagine that we're at the end of a particularly warm week of October, and based on the predictions of the model, the company had increased the stock of the ice cream stands ahead of time. However, the weekly sales, while higher than usual for this week of October, have fallen quite short of the quantity predicted by the model. Oops! What happened? Should the data analyst be fired?

What happened is that the model doesn't account for a crucial fact: most of the sales of ice cream take place during the summer months when kids are out of school. The regression model made its best prediction with the data available, but part of the cause of increased ice cream sales (summer break for students) was misattributed to temperature because summer months are positively correlated with temperature. Since the temperature increase in October did not suddenly make it summer break (sorry, kids!), we saw lower sales than we did on summer days at that temperature.

In technical terms, the month of the year is a confounder of our relationship between temperature and sales. A *confounder* is a variable that introduces bias in a regression; when a confounder is present in the situation you're analyzing, it means that interpreting the regression coefficient as causal will lead to improper conclusions.

Let's think of a place like Chicago, which has a continental climate: winter is very cold and summer is very hot. When comparing sales on a random hot day with sales on a random cold day without accounting for their respective month of the year, you're very likely to be comparing sales on a hot day of summer, when kids are out of school, with sales on a cold day of winter, when kids are in school; this inflates the apparent relationship between temperature and sales.

In this example, we might also expect to see a consistent underprediction of sales in colder weather. In truth, there is a paradigm shift in summer months, and when that shift has to be managed exclusively through temperature in a linear regression, the regression coefficient for temperature will invariably be too high for warmer temperatures and too low for colder temperatures.

## Too Many Variables Can Spoil the Broth

A potential solution to the problem of confounders would be to add to the regression all the variables we can. This mindset of "everything and the kitchen sink" still has proponents among statisticians. In *The Book of Why*, Judea Pearl and Dana Mackenzie mention that "a leading statistician even recently wrote, 'to avoid conditioning on some observed covariates…is nonscientific ad hockery'" (Pearl & Mackenzie 2018, p. 160).[2] It is also quite common among data scientists. To be fair, if your goal is only to predict a variable, you have a model that is carefully designed to generalize adequately beyond your testing data, and you don't care about why the predicted variable is taking a certain value, then that's a perfectly valid stance. But this does not work if your goal is to understand causal relationships in order to act upon them. Because of

---

2 In case you're wondering, the aforementioned statistician is Donald Rubin.

this, just adding as many variables as you can to your model not only is inefficient but can be downright counterproductive and misleading.

Let's demonstrate this with our example by adding a variable that we might be inclined to include but will bias our regression. I created the variable *IcedCoffeeSales* to be correlated with *Temperature* but not with *SummerMonth*. Let's look at what happens to our regression if we add this variable in addition to *Temperature* and *SummerMonth* (a binary 1/0 variable that indicates if the month was July or August (1) or any other month (0)):

```
## R (output not shown)
> summary(lm(icecream_sales ~ iced_coffee_sales + temps + summer_months))

## Python
print(ols("icecream_sales ~ temps + summer_months + iced_coffee_sales",
          data=stand_data_df).fit().summary())
...
                      coef    std err       t    P>|t|    [0.025     0.975]
--------------------------------------------------------------------------
Intercept          -15.8271   374.581   -0.042   0.966   -750.363   718.709
temps             2702.7885  2083.161    1.297   0.195  -1382.196  6787.773
summer_months     1.955e+04   361.572   54.064   0.000   1.88e+04  2.03e+04
iced_coffee_sales   -1.7011     2.083   -0.817   0.414     -5.786     2.383
...
```

We see that the coefficient for *Temperature* has shifted dramatically from our prior example, but in the wrong direction, and it's now further away from the true value. And despite *IceCreamSales* and *IcedCoffeeSales* being positively correlated, the coefficient for the latter is negative. The high p-values for *Temperature* and *IcedCoffeeSales* would usually be taken as signs that something is amiss, but since the p-value for *Temperature* is worse than before, an analyst may conclude that they should remove it from the regression. How is this possible?

The truth behind the data (which is known, since I manufactured the relationships and randomized data around those relationships) is that when it is hot out people are more likely to buy iced coffee. On hot days, people are also more likely to buy more ice cream. But a purchase of iced coffee itself does not make customers any more or less likely to buy ice cream. Summer months are also not correlated with iced coffee purchases since schoolchildren are not a significant factor in the demand for iced coffee (see the sidebar for the details of the math at hand).

# Technical Deeper Dive: What Happened Here?

The equation for ice cream sales that I used to generate the simulated data is:

$$IceCreamSales := 1{,}000.Temperature + 20{,}000.SummerMonth + \varepsilon_1$$

where $\varepsilon_1$ represents some random noise with a mean of zero and the ":=" sign indicates that this equation represents how the variable on the left, *IceCreamSales*, is defined or constructed.

However, the equation we're estimating in our linear regression is:

$$IceCreamSales = \beta_T.Temperature + \beta_S.SummerMonth + \beta_C.IcedCoffeeSales$$

The true equation that was used to generate iced coffee sales is:

$$IcedCoffeeSales := 1{,}000.Temperature + \varepsilon_2$$

Which means we can rewrite the previous equation as:

$$IceCreamSales = \beta_T.Temperature + \beta_S.SummerMonth + \beta_C.(1{,}000.Temperature + \varepsilon_2) = (\beta_T + 1{,}000\ \beta_C).Temperature + \beta_S.SummerMonth$$

Barring some random fluke, our coefficient $\beta_S$ should be close to the true value 20,000. But for temperature, our software will try to solve this equation:

$$\beta_T + 1{,}000.\beta_C = 1{,}000$$

This is a single equation with two unknowns, so it has an infinite number of solutions. $\beta_T = 0$ and $\beta_C = 1$ would work, but so would $\beta_T = 500$ and $\beta_C = 0.5$, or $\beta_T = 5{,}000$ and $\beta_C = -4$. The least-square algorithm will pick the combination that provides the highest $R^2$ value, but it will not be reliable (although the unreliability will generally be much smaller in practice than in this simulated example). In technical terms, we have introduced multicollinearity.

Figure 1-4 shows a positive correlation between iced coffee sales and ice cream sales since both increase when it is warmer out, but any increase in sales of iced coffee during summer months can be explained by a shared correlation with the temperature variable. When the regression algorithm tries to explain ice cream sales using the three variables at hand, the explanatory power of temperature on iced coffee sales was added to the temperature variable while iced coffee was forced to compensate for the overpowering of temperature. Even though iced coffee sales are not statistically significant and the coefficient is relatively small, the dollars of sales are much higher than degrees of temperature, so ultimately iced coffee sales cancel the inflation of the coefficient for temperature.

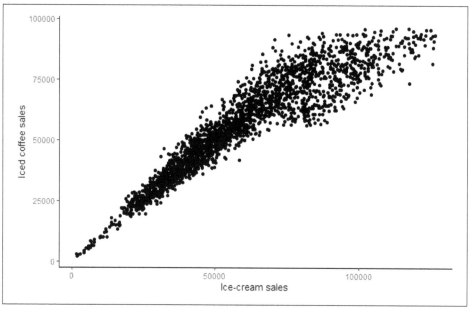

*Figure 1-4. Plot of iced coffee sales versus ice cream sales*

In the previous example, adding the variable *IcedCoffeeSales* to the regression muddled the relationship between temperature and ice cream sales. Unfortunately, the reverse can also be true: including the wrong variable in a regression can create the illusion of a relationship when there is none.

Sticking with our ice cream example at C-Mart, let's say that the category manager is interested in understanding customer tastes, so they ask an employee to stand outside the store and survey people walking by, asking them how much they like vanilla ice cream and how much they like chocolate ice cream (both on a scale from 0 to 25), as well as whether they have ever purchased ice cream from the stand. To keep things simple, we'll assume that the stand only sells chocolate and vanilla ice cream.

Let's assume for the sake of the example that taste for vanilla ice cream and taste for chocolate ice cream are entirely uncorrelated. Some people like one but not the other, some like both equally, some like one *more than* the other, and so on. But all of these preferences impact whether someone buys from the stand, a binary (Yes/No) variable.

Because the variable *Shopped* is binary, we would use a logistic regression if we wanted to measure the impact of either of the *Taste* variables on shopping behavior. Since the two *Taste* variables are uncorrelated, we would see a regular cloud with no apparent correlation if we were to plot them against each other; however, they each impact the probability of shopping at the ice cream stand (Figure 1-5).

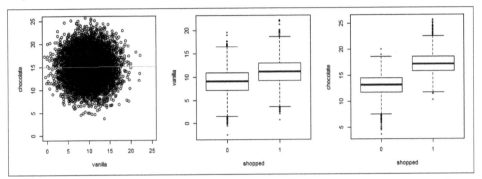

*Figure 1-5. Left panel: tastes for vanilla and chocolate are uncorrelated in the overall population; middle panel: taste for vanilla is higher for people who shop at the ice cream stand than for people who don't; right panel: same result for taste for chocolate*

In the first graph, I added a line of best fit, which is almost perfectly flat, reflecting the lack of correlation between the variables (the correlation coefficient is equal to 0.004, reflecting sampling error). On the second and third graphs, we can see that taste for vanilla and chocolate is higher on average for customers (*Shopped* = 1) than for non-customers, which makes sense.

So far, so good. Let's say that once you get the survey data, your business partner tells you that they are considering introducing a coupon incentive for the ice cream stand: when you purchase ice cream, you get a coupon for future visits. This loyalty incentive won't impact the respondents who have never shopped at the stand, so the relevant population are those who have shopped at the store. The business partner is considering using flavor restrictions on the coupons to balance stock but does not know how much flavor choices can be influenced. If someone who purchased vanilla ice cream were given a coupon for 50% off chocolate ice cream, would it do anything beyond adding more paper to the recycle bin? How favorably do the people who like vanilla ice cream view chocolate ice cream anyway?

You plot the same graph again, this time restricting the data to people who have answered "Yes" to the shopping question (Figure 1-6).

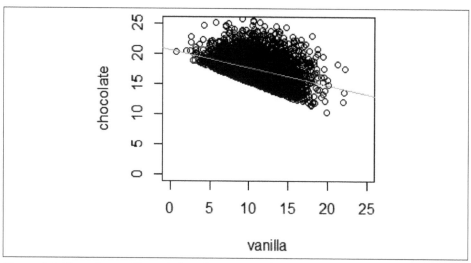

*Figure 1-6. Taste for vanilla and chocolate among shoppers*

There is now a strong negative correlation between the two variables (the correlation coefficient is equal to −0.39). What happened? Do vanilla lovers who come to your stand turn into chocolate haters and vice versa? Of course not. This correlation was artificially created when you restrained yourself to customers.

Let's get back to our true causal relationships: the stronger someone's taste for vanilla, the more likely they are to shop at your stand, and similarly for chocolate. This means that there is a cumulative effect of these two variables. If someone has a weak taste for both vanilla and chocolate ice creams, they are very unlikely to shop at your stand; in other words, most of the people with a weak taste for vanilla among your customers have a strong taste for chocolate. On the other hand, if someone has a strong taste for vanilla, they might shop at your stand even if they don't have a strong taste for chocolate. You can see it reflected in the earlier graph: for high values for vanilla (say above 15), there are data points with lower values for chocolate (below 15), whereas, for low values of vanilla (below 5), the only data points in the graph have a high value for chocolate (above 17). No one's preferences have changed, but people with a weak taste for both vanilla and chocolate are excluded from your data set.

The technical term for this phenomenon is the Berkson paradox (*https://oreil.ly/KwJ1R*), but Judea Pearl and Dana Mackenzie call it by a more intuitive name: the "explain-away effect." If one of your customers has a strong taste for vanilla, this completely explains why they are shopping at your stand, and they don't "need" to have a strong taste for chocolate. On the other hand, if one of your customers has a weak taste for vanilla, this can't explain why they are shopping at your stand, and they must have a stronger than average taste for chocolate.

The Berkson paradox is counterintuitive and hard to understand at first. It can cause biases in your data, depending on how it was collected, even before you start any analysis. A classic example of how this situation can create artificial correlations is that some diseases show a higher degree of correlation when looking at the population of hospital patients compared to the general population. In reality of course, what happens is that either disease is not enough for someone to go to a hospital; someone's health status gets bad enough to justify hospitalization only when they are both present.[3]

## Conclusion

Predictive analytics has been extremely successful over the past few decades and will remain so. On the other hand, when trying to understand and—more importantly— change human behavior, causal analytics offers a compelling alternative.

Causal analytics, however, demands a different approach than what we are used to with predictive analytics. Hopefully, the examples in this chapter have convinced you that you can't just throw a bunch of variables in a linear or logistic regression and hope for the best (which we might think of as the "include them all, God will recognize His own" approach). You may still wonder, though, about other types of models and algorithms. Are gradient boosting or deep learning models somehow immune to confounders, multicollinearity, and spurious correlations? Unfortunately, the answer is no. If anything, the "black box" nature of these models means that confounders can be harder to catch.

In the next chapter, we will explore how to think about the behavioral data itself.

---

3 Technically speaking, this is a slightly different situation, because there is a threshold effect instead of two linear (or logistic) relationships, but the underlying principle that including the wrong variable can create artificial correlations still applies.

# CHAPTER 2

# Understanding Behavioral Data

In Chapter 1 we discussed that the core goal of this book is to use data analysis in order to understand what drives behavior. This requires understanding the relationship between data and behaviors, which was represented in Chapter 1 by an arrow in the causal-behavioral framework (Figure 2-1).

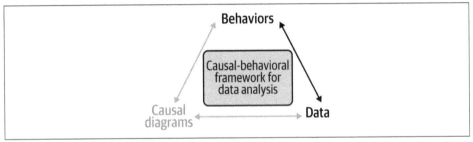

*Figure 2-1. The causal-behavioral framework with the arrow for this chapter highlighted*

Forgive the pop-culture reference, but if you've seen *The Matrix*, you'll recall that the main character can look at the world around him and see the numbers behind it. Well, in this chapter you'll learn to look at your data and see the behaviors behind it.

The first section is mostly directed to readers with a business or data analysis background and offers a basic "behavioral science 101" introduction to core behavioral science concepts. If you're a behavioral scientist by training, you probably won't find much that's new here, but you may want to skim through just to get a feel for the specific terms I use.

Based on that common understanding, the second section will show you how to look at your data through a behavioral lens and determine the behavioral concept related to each variable. In many cases, unfortunately, a variable is at first only loosely related to the corresponding behavior, so we'll also learn how to "behavioralize" such wayward variables.

# A Basic Model of Human Behavior

"Behavior" is one of those words that is highly familiar to us due to repeated exposure but is rarely, if ever, properly defined. I once asked a business partner what behavior she was trying to encourage, and her answer started with "we want them to know that…" In that moment, I realized two things: (1) by helping clarify the goals at hand, I could add more value to that project than I initially expected, and (2) the introduction to behavioral science I had walked her through earlier really sucked if she still thought that knowing something was a behavior. Hopefully I'll do a better job this time and you'll come out of this section able to experience the joy of adding more value to your organization.

Indeed, I strongly believe that one of the key benefits of a behavioral mindset is to get people to think more precisely about what it is they're trying to do. Changing someone's mind is not the same thing as influencing their actions, and vice versa. For that purpose, I'll propose a simplified but hopefully actionable model of human behavior, which I'll first illustrate with an example of how a personal care company could capitalize on a customer's midlife crisis (Figure 2-2).

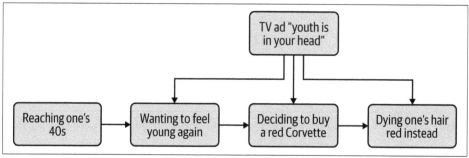

*Figure 2-2. Our model of human behavior for a midlife crisis*

In this example, personal characteristics (reaching one's 40s) lead to emotions and thoughts (wanting to feel young again), which in turn lead to an intention (deciding to buy a red Corvette). That intention may translate into the corresponding behavior, or it may result in another action (dying one's hair red instead), depending on business behaviors (running a TV ad).

In some circumstances we may be trying to affect not the behaviors of our customers but those of our employees, suppliers, and so on. We would have to adjust this model accordingly, but the intuition would remain the same: on the one hand, there is a human being whose behavior we're trying to influence, and on the other hand, there are all the processes, rules, and decisions that our business controls (Figure 2-3).

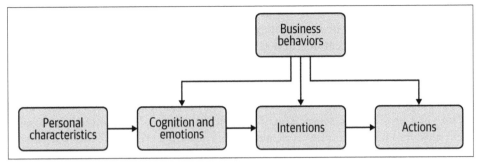

*Figure 2-3. Our general model of human behavior*

In this model, personal characteristics affect cognition and emotions, which in turn affect intentions, which affect actions. Business behaviors, in the form of processes, rules, and decisions that we control, affect all three categories.

There are many different models out there, and there's nothing definitive or magical about this one. But for the data analysis stage, which is the topic of this book, I feel that these five components best correspond to the types of data you'll encounter. Let's look at these one by one, reviewing what defines each component and exploring how we can collect and handle data about each in an ethical manner.

## Personal Characteristics

Collecting and using demographic variables such as age, gender, and family situation is a staple of applied analytics, and for good reason: these can often be good predictors of what someone will do. As behavioral data analysts, however, we'll want to think about personal characteristics both more broadly and more precisely. For our purposes, we'll define as *personal characteristics* all the information we have about a person that changes only rarely or very gradually over the relevant time frame for our analysis.

Getting back to the example of C-Mart's ice cream stands, this means including things like personality traits and lifestyle habits. Let's say that one of your customers has a high degree of openness to experience and is always game to try some new flavor combination like blueberry and cheese; you can reasonably consider that trait as a stable personal characteristic for your analysis, even though psychology tells us that personality changes in predictable ways over someone's lifetime. On the other side of the spectrum, an urban planner looking decades in the future may want to take into account not only gradual changes in lifestyle habits, but also potential divergences whereby the direction of these changes is itself changing. In that sense, I think of personal characteristics as "primary causes" defined to avoid an infinite regression: we track their changes but we disregard their own causes.

While demographic variables fit the preceding definition in terms of relative stability and primacy, it may not be immediately clear what it means to treat them as causes of cognition, emotions, intentions, and actions. Some readers may even feel uncomfortable saying that age or gender is causing someone to do something, as it can sound suspiciously like social determinism.

I would argue that we can resolve these issues by defining a cause as a "contributing factor" in a probabilistic sense. Reaching one's 40s is neither a necessary nor a sufficient reason to have a midlife crisis, and having a midlife crisis is neither a necessary nor a sufficient reason to buy a red Corvette. In fact, both causal relationships are very heavily intertwined with other contributing factors: the contribution of age to existential qualms depends a lot on social patterns such as professional and family trajectories (e.g., the age at which someone enters the labor market or has their first child, assuming they do either one), and resolving such qualms through consumption is possible only if one has enough available income. It is also dependent on the degree to which one is influenced by advertising.

As a behavioral science mantra puts it, "behavior is a function of the person and the environment," and social factors often have arguably more weight than demographic variables. From a causal modeling and data analysis perspective, this interplay between social phenomena and personal characteristics can be captured through the use of moderation and mediation (which we'll explore in Chapters 11 and 12).

In that sense, demographic variables are often useful in behavioral data analysis, because their predictive power hints at the presence of other personal characteristics (of which they are a contributing factor) that are more psychological and actionable in nature. For example, law enforcement and nursing are still overwhelmingly gendered fields in the US and many other countries. Noting that empirical regularity is unlikely to earn you any pat on the back. A much more interesting question is how that effect happens. For instance, one hypothesis would be that it stems from social representations and norms about authority and care. Alternatively, one could posit that this phenomenon is self-perpetuating because of the lack of alternative role models. Determining which hypothesis is true (or which has more weight if both are true) could lead to much more effective career counseling for high-schoolers.

## Collecting data and ethical considerations

Demographic information is widely available and used (one could even say overused) in business analytics. The main challenge is to identify and measure other personal characteristics so that you can correctly assign their effects on behaviors instead of unduly attributing them to demographic variables. Market and user research can be very useful here.

Misattributing effects also matters from an ethical perspective: in many cases, discrimination can be thought of as consciously or unconsciously misattributing behaviors, either current or predicted, to demographic variables. Even when your analysis is correct, you also need to ensure that it doesn't inadvertently lead to reinforcing patterns of advantage and disadvantage.

## Cognition and Emotions

*Cognition and emotions* is a phrase I'm using to encompass mental states such as emotions, thoughts, mental models, and beliefs. You can think of it as everything that happens in a customer's brain apart from intentions and more permanent personal characteristics. For example, is a potential customer aware of your company and its products or services? If so, what do they think about them? What is their mental model of how a bank works? Do they feel a pinch of envy every time they see a sleek new Tesla on the road?

Cognition and emotions encompass all of that, as well as more nebulous business buzzwords and phrases such as *customer satisfaction* (CSAT) and *customer experience* (CX). CX has become a business mantra: many companies have a CX team, and there are conferences, consultants, and books devoted to the topic. But what exactly is it? Can you measure its causes and its effects? Yes, you can, but it requires intellectual humility and the willingness to spend time on sleuthing work.

### Collecting data and ethical considerations

Cognition and emotions are like psychological personal characteristics in two respects: they generally can't be observed directly, and the relevant data is collected through surveys or during user experience (UX) observation. It's important to note that unless you're using physiological measures, you'll be relying on *stated* or *observed* measures.

This brings us to one of the biggest differences between UX or human-centered design and behavioral science: UX begins with the presumption that human beings know what they want, how they feel about something, and why, whereas behavioral science begins with the presumption that we are unaware of a lot of things going on in our own heads. To use a legal metaphor, a behavioral scientist will often treat what someone says as suspect until proven trustworthy, whereas a UX researcher will treat it as honest until proven misleading.

However I don't want to exaggerate the difference, and the distinction often gets blurred in practice, depending on the situation at hand. If a behavioral scientist is told by a customer that a website is confusing and frustrating to use, they will take a page from the UX book and trust that the customer did indeed experience negative emotions. Conversely, when doing foundational product research, a skilled UX researcher will often go beyond stated intentions and attempt to identify a customer's deeper needs.

From an ethical perspective, trying to change someone's cognition and emotions is definitely a gray area. My recommended litmus test in that regard is the "NYT" test: are your intentions and methods benevolent and transparent enough that your boss's boss wouldn't mind seeing them displayed on the front page of the *New York Times*? Advertising strives to affect people's thoughts and emotions, but I don't expect to ever see a headline reading, "Companies spending billions to make people want their stuff"; it would be met by a big yawn anyway. On the other hand, "sludges"—manipulative behavioral science tricks such as "three other people are looking at this property right now!"—and good old-fashioned lies wouldn't pass that test. This still leaves many honest opportunities to apply behavioral science: for example, explaining your product's benefits in terms that resonate better with your customers' mental models or making a purchasing experience easier or more enjoyable. Because the topic of this book is data analysis and not behavioral design, we'll look only at how to analyze such interventions, not how to develop them.

## Intentions

When someone says "I'm going to do X," whether X is buying groceries for the week or booking a vacation, they express an *intention*. As such, intentions bring us one step closer to actions.

Technically speaking, an intention is a mental state that could have been included in the previous category of cognition and emotions, but I felt it deserved its own category because of its importance in applied behavioral data analysis. Unless you plan to force customers down a certain path (e.g., remove the phone number from your website so that they don't call you), you'll generally have to use their intention as a conduit for behavior change.

In addition, people often fail to follow through on things they want to do, a concept known in behavioral science as the intention-action gap. (As an example, think of New Year's resolutions and how often they are broken.) Therefore, a key to driving customer behavior is determining whether a potential response doesn't occur because the customers *don't want to* take that action or because something happens between intention and action.

### Collecting data and ethical considerations

In the general course of business, we usually don't have direct data on intentions. Therefore most of this data will come from customer statements, generally via two sources:

*Surveys*

> These can be either "in the moment" (e.g., at the end of a website or store visit) or asynchronous (e.g., administered by mail, email, or phone)

*User experience observations*

> In these user experience observations a UX researcher asks a subject to go through a certain experience and collects information about their intentions (either by asking them to think aloud during the process or by asking questions afterward, such as, "What were you trying to do at that point?").

Alternatively, we'll often try to infer intentions from observed behaviors, which is known in analytics as "intent modeling."

From an ethical perspective, influencing someone's intentions is the objective of advertising, marketing, and the dark arts of persuasion. It can be done, and I genuinely believe it can even be done ethically in a number of situations, as UX and behavioral researchers strive to do. But the simple truth is that it's generally hard to do and it can backfire. It is much simpler and safer to identify spots where you can help customers bridge the intention-action gap. To recast a common business phrase in our model, a "pain point" often reflects an obstacle to achieving an intent.

## Actions

An *action* is the basic unit of a *behavior*, and I'll often use these two words interchangeably. The rule of thumb I often give people is that an action or behavior is something you should be able to observe if you were in the room at that moment without having to ask the person. "Buying something on Amazon" is an action. So is "reading a review of a product on Amazon." But "knowing something" or "deciding to buy something on Amazon" is not. You can't know that someone has made a decision unless you either ask them or see them acting on that decision (which is a consequence but not the same thing).

An action or behavior can often be defined at various levels of granularity. "Going to the gym" is observable, and in many situations that's an acceptable level of detail. If you're sending people a coupon for a free visit and they show up, you've achieved your goal, and there is no need to think more deeply. However, if you're working on a wellness app and you want to make sure that people stay on track with their program, you'll need to be much more granular than that. I once attended a presentation by behavioral scientist Steve Wendel in which he walked the audience through "going to the gym" step by step (putting on gym clothes, going to the car, deciding what exercise to do once at the gym, and so on). This is the level of detail you'll often need to employ if you are attempting to identify and resolve behavioral blockers. The same would apply to a customer journey on a website or in an app. What does "signing up" mean? What information do they need to retrieve and enter? What choices are you asking them to make?

### Collecting data and ethical considerations

Action or behavior data often represents the largest category of available customer data and generally falls under the heading of "transactional data." In many circumstances, however, this data will need some additional processing to truly reflect behaviors. We'll look more closely at this in the second section of this chapter.

If you're targeting employee behavior (e.g., in people analytics or to reduce attrition), data may be much sparser and harder to access, as it will often need to be pulled from business process software and may not be recorded as a matter of course.

Let's call a spade a spade: modifying behavior is the ultimate goal of behavioral analytics. In some instances, it can be very simple: if you remove your customer service phone number from your website, fewer people will call you. Whether or not that's a good idea is a different story. And obviously, behavior modification can sometimes be extremely difficult, as illustrated by the large and growing literature on behavior change.[1]

Either way, the goal of modifying behavior comes with ethical responsibilities. Here again, the NYT question is my recommended litmus test.

## Business Behaviors

The last category of data we need to account for is that of *business behaviors*: things an organization or one of its employees do that affect a customer (or another employee if your focus is on employee behavior). This includes:

- Communication such as emails and letters
- Changes to website language or in the talk path for a call center representative
- Business rules, such as the criteria for customer rewards (e.g., "if a customer has spent X dollars in the last six months, send them a coupon") or for hiring decisions
- Decisions by individual employees, such as labeling a customer account as potentially fraudulent or promoting another employee

---

1 Thaler and Sunstein's *Nudge* (2009) and Eyal's *Hooked: How to Build Habit-Forming Products* (2014) are two good references among many on the topic.

## Collecting data

Business behaviors can be both a blessing and a curse to behavioral data analysts. On the one hand, from an intervention design perspective, they're one of our main levers to drive individual behaviors: for example, we could modify the content of a letter or the frequency of an email to persuade a customer to pay their bill on time. As such, business behaviors are an extremely valuable tool without which our job would be significantly harder, and when we are the ones making changes to business behaviors, it is generally reasonably easy to collect the corresponding data. Another advantage of business behaviors is that there are fewer ethical considerations about data collection: organizations are generally free to collect data about their own operations and their employees' behavior in the normal course of business (although there are certainly circumstances where it can be considered an invasion of privacy).

On the other hand, from a data collection and analysis perspective, business behaviors can be an analyst's worst nightmare: like water to fish, they can be invisible to an organization, and their effects on individual behaviors then become intractable noise. This happens for two reasons.

First, many organizations, if they track business behaviors at all, simply don't track them at the same level of detail as customer behaviors. Let's say that C-Mart experimented with reducing its hours of business in the summer of 2018, leading to a temporary reduction in sales. Good luck figuring that out from the data alone! Many business rules, even when they are implemented in software, are simply not logged anywhere in a machine-readable format. If the corresponding data has indeed been recorded, it is often stored only in a departmental database (or worse, an Excel file) instead of the enterprise data lake.

Second, business behaviors can affect the interpretation of variables for customer behaviors. The clearest example of that would be *sludges*—intentional frictions and misleading communication introduced to confuse customers. Imagine a form on a website which, when you enter your email address, automatically checks the box "I want to receive marketing emails" that you had unchecked at the beginning of the form. Would that checked box really indicate that the customer wants to receive marketing emails? Beyond such obvious examples, business behaviors can be found lurking behind many customer behaviors, especially in the realm of sales. Many propensity-to-buy models should have as a caveat "among the people our sales team decided to call." Paradoxically, while the compensation structure of sales representatives is often one of the levers that business leaders obsess most about, it is rarely included in models of customer purchasing behaviors.

Ultimately, getting reliable data about business behaviors, especially over time, can be a formidable challenge for behavioral data analysts—but that means that it's also one way they can create value for their organization before running any analysis.

# How to Connect Behaviors and Data

Personal characteristics, cognition and emotions, intentions, actions, and business behaviors: these are the concepts we have at our disposal to represent and understand our world as behavioral data analysts. However, connecting behaviors and data is not simply a matter of assigning the variables at hand to one of these buckets. Being "about behaviors" is not enough for a variable to qualify as "behavioral data"; as we saw in the previous section, for instance, a variable supposedly about customer behavior may really only reflect a business rule. In this section, I'll give you a list of tips to behavioralize your data and ensure that it fits as closely as possible the qualitative reality it's supposed to represent.

To make things more concrete, we'll look at our second fictional company, Air Coach and Couch (AirCnC), an online travel-and-lodging booking company. AirCnC's leadership has asked their analyst to measure the effect of CSAT on future purchasing behavior, namely the amount spent in the six months following a booking (*M6Spend*), one of their key performance indicators. We'll answer that common but intractable business question by the end of the book, in Chapter 12. For now, we'll just see how to get started with framing it.

## Develop a Behavioral Integrity Mindset

Because behavioral science is so new to businesses, you'll generally be the first person to bring that lens to your organization's data, which will likely contain dozens, if not hundreds or even thousands, of variables. That's a daunting task, but adopting the right mindset will help you make sense of it and get started.

Imagine for a second that you're a structural engineer who has been newly assigned the responsibility of maintaining a bridge. You could start on one end of it and assess its integrity inch by inch (or centimeter by centimeter) until you've reached the other end and then come up with a 10-year plan to reach perfect structural integrity. Meanwhile potholes are getting worse, endangering automobilists and damaging their vehicles trip after trip. A sounder approach would be to take a quick look around, prioritize major problems that can be fixed quickly, and identify places where temporary expedients will do until you have the time and budget to make more structural changes.[2]

The same mindset applies to your data. Unless you happen to be among the very first employees of a startup, you'll be dealing with existing data and legacy processes. Don't panic, and don't start going through your table list in alphabetical order. Start

---

2 At this point, I would like to preemptively apologize to structural engineers if, as is likely, this metaphor gravely misrepresents their work. Poetic license and all that.

with a specific business problem and identify the variables that are most likely to be inaccurate, in decreasing order of their importance for the business problem:

1. Causes and effects of interest

2. Mediators and moderators, if relevant

3. Any potential confounder

4. Other nonconfounding independent variables (a.k.a. covariates)

You'll have to make judgment calls along the way: for example, should you include a certain variable in your analysis, or is it so poorly defined that you're better off without it? Unfortunately, there is no clear-cut criterion to make these calls correctly; you'll have to rely on your business sense and expertise. There is, however, a clear-cut way to make these calls *incorrectly*: pretend that they don't exist. A variable will or will not be included in your analysis, and there is no way around that fact. If your instinct leans toward inclusion, as is likely, then document why, describe potential sources of error, and indicate how the results would be different if the variable were omitted. As a UX researcher once put it in a friendly chat with me, being a researcher in business means constantly figuring out "what you can get away with."

## Distrust and Verify

Unfortunately, in many circumstances, the way data is recorded is driven by business and financial rules and is transaction-centric rather than customer-centric. This means that you should consider variables suspicious until proven innocent: in other words, do not automatically assume that the variable *CustomerDidX* means that the customer did X. It may mean something entirely different. For example:

- The customer checked a box without reading the fine print that mentioned that they were agreeing to X.
- The customer didn't say anything, so we defaulted them to X.
- The customer stated that they did X, but we can't verify.
- We bought data from a vendor indicating that the customer regularly did X at some point in their life.

Even if the customer actually did X, we can't assume their intent. They may have done this:

- Because we sent them a reminder email
- Four times in a row because the page was not refreshing
- Mistakenly, when they really wanted to do Y

- A week ago, but due to regulatory constraints we recorded it only today

In other words, to paraphrase a popular line from *The Princess Bride*: "You keep using that variable. I do not think it means what you think it means."

## Identify the Category

As stated, AirCnC's leadership has tasked its analysts with understanding the effects of CSAT on purchasing behaviors, which is a tall order. Our first step is to figure out what we're talking about.

In my first year of college, my philosophy professor would give us essay prompts such as "What is progress?" and "Man and the Machine." Along with them, he offered some excellent advice if we got stumped: we should figure out, in his words, "what book this is a chapter of." Paradoxically, when a question seems intimidatingly big and intractable, it often helps to ask what broader category it belongs to.

Fortunately, as behavioral data analysts, we don't need to roam the Library of Congress for inspiration; we have the tailor-made categorization from earlier in the chapter:

- Personal characteristics
- Cognition and emotions
- Intentions
- Actions (a.k.a. behaviors)
- Business behaviors and processes

Let's proceed by elimination. Customer satisfaction is not fixed and permanent, so it's not part of personal characteristics. It's neither something people do, nor something people intend to do, so it's not a behavior or an intention. Finally, it's not something happening purely on the business side, so it's not a business behavior or process. Therefore, customer satisfaction is part of cognitions and emotions, as is its cousin customer experience. The second variable in our business problem, "purchasing behaviors," is much easier to categorize: it's obviously a customer behavior.

In my experience, many business analytics projects fail or deliver underwhelming results because the analyst has not clarified what the project is about. Organizations always have an overarching target metric—profit for companies, client outcomes for nonprofits, etc. At a lower level, departments often have their own target metrics, such as Net Promoter Score for the customer experience team, downtime percent for IT, etc. If a business partner asks you to measure or improve a variable that seems unrelated to one of these target metrics, it generally means that they have in mind an implicit and possibly faulty behavioral theory connecting the two.

"Customer engagement," another buzzword concept that behavioral scientists are often asked to improve, is a good example of that phenomenon. It's not clear cut where it belongs, because it could really refer to two different things:

- A behavior, namely the broad pattern of interactions with the business: customer A is deemed more engaged than customer B if customer A logs on to the website more often and spends more time navigating it.
- A cognition or emotion, as when an audience is "engaged" with a movie or a course because they are engrossed in the flow and eager to know what comes next.

Indeed, I strongly believe that the confusion between these two things explains a large part of the appeal of engagement metrics for startups and the broader digital world, even though they may be misleading. For example, in the first sense of the word, I'm more engaged with my washing machine when it stops working; that doesn't translate into enjoyment and eagerness in the second sense of the word. Organizations that try to increase engagement as a behavior are often disappointed with the results. When engagement as behavior doesn't translate into engagement as emotion, it doesn't lead to desirable outcomes such as higher loyalty and retention.

As a personal example, a business partner once asked me for help to get employees to do a certain training. After some discussion, it became clear that what she really wanted was for employees to comply with a business rule; she believed that they didn't comply because they were not sufficiently informed about the rule. We pivoted the project toward understanding why employees didn't comply and how to encourage them to do so. In short: beware the self-diagnosing patients!

We can now recast our business problem in our model of human behavior: AirCnC leaders want to know whether a cognition/emotion affects a behavior in customers, and if so, how much. Indeed, in the overwhelming majority of business analytics problems, at least one of the variables involved is a behavior, either of a customer or of the business. Are customers more satisfied if we send them a follow-up email? Are satisfied customers more likely to buy?

If, after identifying the categories of the relevant variables, none of them falls under the customer or business behavior categories, it should raise alarms. This suggests a business problem without a clear "So what?" Let's say that older customers are more satisfied. So what? What are we going to do with that information? Business outcomes are driven by behaviors, either ours or those of our customers.

Once you have identified the relevant behaviors, it's time to drill down on the corresponding variables.

# Refine Behavioral Variables

As I mentioned earlier, a variable being "about a behavior" is not the same thing as being a behavioral variable. You'll often have to transform them to make them truly behavioral.

Let's focus on customer behaviors, as they are more intuitive, but the logic applies similarly to business behaviors. A good behavioral variable will have the following characteristics:

*Observable*

As mentioned in Chapter 1, a behavior is observable, at least in principle. If you were in the room with the customer, could you *see* them do it? Abandoning a booking in the middle of the process is observable; changing one's mind is not. A good clue is that if it doesn't have a timestamp, it's probably not concrete and granular enough.

*Individual*

Businesses often rely on aggregate metrics such as rates or proportions (e.g., proportion of customers who cancel their account). Aggregate metrics can offer useful snapshots for reporting purposes, but they can fall prey to biases and confounding factors such as changes in population composition (a.k.a. customer mix), especially when they are calculated based on time intervals.

For example, let's imagine that a successful marketing campaign brings a lot of new users to AirCnC's website. Let's also assume that in that line of business, a significant share of new customers cancel their account in their first month. Thus AirCnC's daily cancel rate may spike alarmingly during the month following the campaign, even though nothing went wrong. A good rule of thumb is that sound aggregate variables are based on sound individual variables. If a variable makes sense only in the aggregate and doesn't have a meaningful interpretation at the individual level, that's a red flag. In our example, a meaningful individual counterpart to the cancellation rate would be the cancellation probability. When controlling for individual characteristics and tenure with the company, this metric would remain stable despite the influx of new customers.

*Atomic*

Similarly, businesses often aggregate together a variety of different behaviors sharing a common intent. For example, there might be three different ways for a customer to change their billing address with AirCnC: by going to their account settings, by editing the information when finalizing a booking, and by contacting the call center. These would look different to someone watching the customer in the moment, but may be logged similarly in the database. Again, I don't want to suggest that aggregate behavioral variables are inherently bad. There are certainly analyses that call for using the binary variable *ChangedBillingInformation*. But at the very least, you should be aware of the concrete ways it can be done, and it doesn't hurt to check whenever possible that the overall conclusion you reached applies to each one of them.

In many circumstances, identifying or creating a satisfying behavioral variable involves "getting your hands dirty." Databases for analytical or research purposes often offer a "cleaned-up" version of the truth as it would appear in the transaction databases, listing only the most up-to-date, vetted-out information. This makes perfect sense in most circumstances: if a customer made a booking and then canceled and was refunded, we wouldn't want that amount to count toward the *AmountSpent* variable. After all, from a business perspective, AirCnC didn't get to keep that money. However, from a behavioral perspective, that customer is different from a customer who didn't make any booking over the same time period, and there are analyses for which it would be relevant to take it into account. Don't go and learn an ancient programming language like COBOL just to access the lowest level databases, but it's worth digging around a bit beyond your usual pretty tables.

## Understand the Context

To reiterate, "behavior is a function of the person and the environment" is a fundamental principle in behavioral science. While personal variables are certainly important, I feel that analysts often rely too much on them because they are readily available, and as a result, they may not think enough about contextual variables.

Often the best way to understand the context(s) in which people behave is through qualitative research such as interviews and surveys, whose insights can be used to generate new variables. Here, I'll focus on how we can tease contextual information out of existing data:

*It's about time*
> As mentioned earlier, a behavior is observable. You can pinpoint the moment in time when it happened, at least in theory and often in practice, thanks to timestamps. Timestamps are gold nuggets for behavioral analysts because they provide intuitive and often readily actionable insights. Unsurprisingly, there's no ready-made algorithm to extract these, only rules of thumb reliant on business sense and the specificities of the problem at hand. I'll give you the most common cues to look for:

*Frequency*
> A natural inclination with behavioral and, more broadly, event data is to look at *frequency*, the count of events/behaviors per unit of time. Unfortunately, frequency data can sometimes misbehave and exhibit artificial discontinuities that do not reflect changes in behaviors. For example, let's say that an AirCnC customer takes a vacation every summer and every winter, which translates into two bookings per year. However, one year they take their

winter vacation in January instead of December. We would count one vacation in the earlier year and three in the latter even though behaviors didn't really change. Such catch-up phenomena—longer durations followed by shorter ones, are best understood by tracking durations directly. For other behaviors, bygones are bygones, and frequency will be a more robust metric.

*Duration*

Duration also offers a natural way to measure decaying effects. Things that you did or that happened a long time ago tend to have smaller effects than more recent occurrences. This often makes duration a good predictive variable. If a customer hasn't left AirCnC after a bad experience five years ago it probably doesn't impact their decisions much anymore, and it would be better to weight the CSAT of past trips by how long ago they happened rather than just use an average.

*Contiguity*

Similarly, behaviors that are very close to each other are often no coincidence and can provide clues to what's going on. A customer calling AirCnC's call center to change their billing information *after* having tried to change it online exhibits a different behavior compared to a customer who calls directly. Siloed data makes omnichannel analytics projects a daunting task, but managing to bring together data from different channels will pay large dividends. One of the best ways to aggregate behavioral data is to create variables for "doing Z after doing X."

*Social schedules*

People are more likely to be at work, or on their way to or from it, Monday through Friday during the day.[3] A customer leisurely scrolling through holiday destinations on a Saturday morning may be at their kid's sport practice. Modern life has its rhythms and schedules that are common knowledge. Because of their granularity, it is often better to start with an "hour of the week" variable instead of having separate "hour of the day" and "day of the week" variables (in local time, of course). Depending on your line of business, you may be able to aggregate things further into variables like "weekday evenings," etc.

*Information and "known unknowns"*

If a tree falls in a forest and it's recorded in your data, but no customer heard it, did it make a sound? Your organization and your customers often know things at different points in time. Variables used in relation to a behavior should always reflect the information available to the person doing it while they're doing it. This

---

3 Unless there happens to be a global pandemic, of course.

can be as simple as replacing "date sent" with "expected date received" for mail you're sending your customer, and vice versa for the mail they send you. People don't know the content of emails they haven't opened. As always, this is part common sense and part behavioral logic, but it will help ensure that your variables snugly fit the corresponding behaviors.

*The dog that didn't bark*

Sometimes people do one thing instead of another, and sometimes they don't have a choice (or they don't see one). What people *don't* do can often be as interesting as what they do. One way to identify alternative behaviors is to look for forks: if behavior B happens often after behavior A, what other behavior C also happens often after A? Conversely, if behavior D often happens after behavior B, what other behaviors lead to behavior D? This can allow you to identify "happy paths" or customers getting lost on their way to getting a job done.

Better understanding the context for a behavior is one of the ways the behavioral perspective brings value to business analytics projects like AirCnC's. Does AirCnC's leadership want to measure customers' satisfaction with their online booking experience, with their stay, or with AirCnC's service representatives (among many other possibilities)? You can ask a customer how satisfied they are at any point starting right after their very first interaction with AirCnC and they'll give you an answer, but it doesn't guarantee that the answers mean the same thing, or even that they mean anything.

In our case, what AirCnC's leadership really wants is to understand how much customer satisfaction after a service call matters. This will help them determine if they should invest more in hiring and training high-quality representatives, or alternatively if they can outsource service to a country with lower wages (hint: they probably shouldn't).

# Conclusion

One of the joys and challenges of applied behavioral science is that it involves a constant back and forth between qualitative and quantitative analysis. My intent in this chapter has been to arm you with a basic model of human behavior and actionable tips so that you can start flexing that muscle and connecting behaviors to data. This means that even before running any data analysis, you can add value to your organization by improving the behavioral integrity of its data and clarifying how to approach business problems.

In the next part of the book, we'll introduce the third pillar of the causal-behavioral framework, namely causal diagrams that will allow us to build relationships between behaviors.

PART II

# Causal Diagrams and Deconfounding

In Part I, we saw how confounding can jeopardize even the simplest data analyses. In Part II, we'll learn to build causal diagrams (CDs) to represent, understand, and deconfound relationships between variables.

First, Chapter 3 provides an introduction to CDs and their building blocks.

In Chapter 4, we'll see how to build a CD from scratch for a new analysis. The CD we saw in our ice cream example was very simple by design. But in real life, it can often be complicated to know what variables to include in our CD beyond our cause and effect of interest, and how to determine the relationships between them.

Similarly, removing confounding from our ice cream example was simple: we just needed to include in our regression a joint cause of our variables of interest. With more complex CDs, it can become difficult to know which variables we should include in our regression. In Chapter 5, we'll see rules that you can apply to even the most complex CD.

# CHAPTER 3

# Introduction to Causal Diagrams

*In fact, with few exceptions, correlation does imply causation. If we observe a systematic relationship between two variables, and we have ruled out the likelihood that this is simply due to a random coincidence, then something must be causing this relationship. When the audience at a Malay shadow theatre sees a solid round shadow on the screen they know that some three-dimensional object has cast it, though they may not know if the object is a ball or a rice bowl in profile. A more accurate sound bite for introductory statistics would be that a simple correlation implies an unresolved causal structure.*

—Bill Shipley, *Cause and Correlation in Biology* (2016)

Causal diagrams (CDs) may well be one of the most powerful tools for analysis most people have never heard of. As such they are one of the three extremities (vertices) of the causal-behavioral framework (Figure 3-1). They provide a language to express and analyze cause-to-effect relationships, which works especially well when dealing with behavioral data analyses.

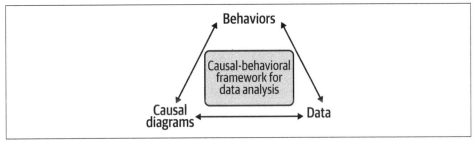

*Figure 3-1. The causal-behavioral framework for data analysis*

In the first section of this chapter, I'll show how CDs fit into the framework from a conceptual perspective, that is, how they are connected to behaviors and data. In the second section, I'll describe the three fundamental structures in CDs: chains, forks, and colliders. Finally, in the third section, we'll see some common transformations that can be applied to CDs.

## Causal Diagrams and the Causal-Behavioral Framework

First, let's define what a CD is. A CD is a visual representation of variables, shown as boxes, and their relationships to each other shown as arrows going from one box to another.

In our C-Mart example in Chapter 1, the variable *IcedCoffeeSales* was affected by a single cause, *Temperature*. Figure 3-2 shows the corresponding causal diagram.

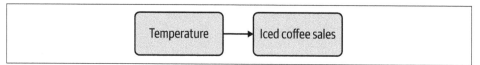

*Figure 3-2. Our very first causal diagram*

Each rectangle represents a variable we can observe (one we have in our data set), and the arrow between them represents the existence and direction of a causal relationship. Here, the arrow between *Temperature* and *IcedCoffeeSales* indicates that the former is a cause of the latter.

Sometimes, however, there will be an additional variable that we aren't able to observe. If we still want to show it in a causal diagram, we can represent it with a shaded rectangle[1] (Figure 3-3).

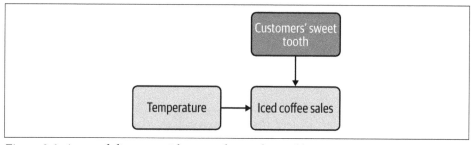

*Figure 3-3. A causal diagram with an unobserved variable*

In Figure 3-3, *CustomerSweetTooth* is a cause of *IcedCoffeeSales*, meaning that customers with a stronger sweet tooth buy more iced coffee. However, we can't observe the degree of a customer's sweet tooth. We'll discuss later the importance of unobserved confounders and more generally unobserved variables in causal analysis. For

---

1 The most common way to represent unobserved variables in CDs is with ovals instead of rectangles.

the time being, let's just note that even if we have no way of observing a particular variable, it can still be included in a causal diagram by representing it as an oval.

## Causal Diagrams Represent Behaviors

The first way of looking at causal diagrams is to treat them as representations of causal relationships between behaviors, as well as other phenomena in the real world that impact behaviors (Figure 3-4). From this perspective, the elements of CDs represent real "things" that exist and have effects on each other. An analogy from physical sciences would be a magnet, a bar of iron, and the magnetic field around the magnet. You can't see the magnetic field but it exists nonetheless, and it affects the iron bar. You may not have any data on the magnetic field and maybe you've never seen the equations describing it, but you can sense it as you move the bar, and you can develop intuitions as to what it does.

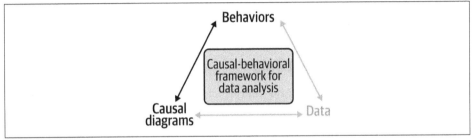

*Figure 3-4. CDs are connected to behaviors in our framework*

The same perspective applies when we want to understand what drives behaviors. We intuitively understand that human beings have habits, preferences, and emotions, and we treat these as causes even though we often don't have any numeric data about them. When we say, "Joe bought peanuts because he was hungry," we are relying on our knowledge, experience, and beliefs about humans in general and Joe in particular. We treat hunger as a real thing, even if we're not measuring Joe's blood sugar or brain activation.

Here, we're making a causal statement about reality: we're saying that had Joe not been hungry, he wouldn't have bought peanuts. Causality is so fundamental to our intuitive understanding of reality that even young children are able to make correct causal inferences (evidenced by their use of the word "because") long before they have had any exposure to the scientific method or data analysis. Of course, intuition is subject to a variety of biases well known by behavioral scientists, even when it takes the more educated form of common sense or expertise. But more often than not, intuition guides us well in our daily lives even in the absence of quantitative data.

You might worry that using CDs to represent intuitions and beliefs about the world introduces subjectivity, and that's certainly true. But because CDs are tools for thinking and analysis, they don't have to be "true." You and I might have different ideas as to why Joe bought peanuts, which means we would draw different CDs. Even if we fully agreed on what causes what, we couldn't represent everything and their relationships in one diagram; there is judgment involved in determining what variables and relationships to include or exclude. In some cases, when data is available, it will help: we'll be able to reject a CD because it is incompatible with the data at hand. But in other cases, very different CDs will be equally compatible with the data and we won't be able to choose between them, especially if we don't have experimental data.

This subjectivity might look like a (possibly fatal) flaw of CDs, but it's actually a feature, not a bug. CDs don't create uncertainty; they simply reflect the uncertainty that is already in our world. If there are several possible interpretations of the situation at hand that appear equally valid, you should explicitly say so. The alternative would be to allow people who have different mental models in their heads to each believe that they know the truth and others agree with them, when in reality that's not the case. At least putting the uncertainty in the open will allow a principled discussion and guide your analysis.

## Causal Diagrams Represent Data

While there is an art to building and interpreting CDs, there's also a science to it, and we can use CDs to represent relationships between variables in our data (Figure 3-5). When these relationships are entirely linear, or approximately so, CDs have clear equivalents in linear algebra. This means we can use the rules and tools of linear algebra to validate the "legality" of how we manipulate and transform CDs, thus ensuring that we draw correct conclusions.

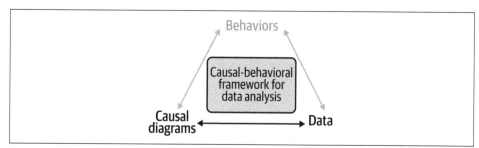

*Figure 3-5. CDs are also connected to data*

The linearity requirement may seem very restrictive. However, some of the rules and tools of linear algebra continue to apply when some of these relationships are not linear but still belong to the broad category of models called generalized linear models (GLM). A logistic regression model for example is a GLM. This means that we can represent and handle a causal relationship where the effect variable is binary with CDs. As the sidebar shows, the math gets more convoluted in that case, but most of our intuitions about CDs remain true.

# Technical Deeper Dive: CDs and Logistic Regression

All the relationships we've seen so far between variables are linear. When the dependent variable is binary (i.e., it can take only two values, usually "Yes" and "No"), we use a logistic regression rather than a linear one. Logistic regression is an example of a GLM and as such it's not linear, but it has some linear characteristics. This implies that the logic of causal diagrams still works, with some hand waving. If you're interested, let's take a look under the hood to make sure you understand what's going on.

Let's look at one of our previous examples, in which taste for vanilla and taste for chocolate cause the purchase of ice cream. This time we'll represent purchases of ice cream as a binary variable: 1 if the customer purchased at least one ice cream that day, 0 if they did not (Figure 3-6).

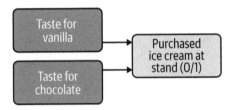

*Figure 3-6. Representing purchases of ice cream as a binary variable in a CD*

Instead of the linear regression we had previously, the arrows in this CD now represent a logistic regression, meaning that the probability that our target variable equals 1 is a transformation of the linear combination of the explanatory variables:

$$P(IceCreamPurchase = 1) = (1 + e^{-(\beta_0 + \beta_V.TasteForVanilla + \beta_C.TasteForChocolate)})^{-1}$$

$F(x) = 1/(1 + e^{-x})$ is called the logistic function, and it results in an S-curve with values between zero and one. The previous equation can therefore be rewritten by inserting the $f$ function into it:

$$P(IceCreamPurchase = 1) = f(\beta_0 + \beta_V.TasteForVanilla + \beta_C.TasteForChocolate)$$

This means that we can't directly translate an increase of 1 "unit" of taste for vanilla ice cream into a fixed increase in the probability of purchasing ice cream. Coefficients of a logistic regression are notoriously hard to interpret for this reason. However, the relationships between the coefficients and the variables within the logistic function remain linear, which means that the algebraic transformations we'll see in the next section will still be correct.

For example, we could conceptually split *TasteForChocolate* into variables representing the taste for various chocolate flavors. The preceding equation would then become:

$$P(IceCreamPurchase = 1) = F(\beta_0 + \beta_V . TasteForVanilla + \beta_{DC} . TasteForDarkChocolate + \beta_{FC} . TasteForFudgeChocolate)$$

Within the logistic function, the explanatory variables are still linear and the linear transformations can still be applied. All of that is to say that I'll mostly be referring to linear regression because it's more common, but you can rest assured that everything I'm saying will also apply to logistic regression, plus or minus some minor math transformation.

From this perspective, the causal diagram from Figure 3-3 connecting *Temperature* to *IcedCoffeeSales* would mean that:

$$IcedCoffeeSales = \beta * Temperature + \varepsilon$$

This linear regression means that if temperature were to increase by one degree, keeping everything else equal, then sales of iced coffee would increase by $\beta$ dollars. Each box in the causal diagram represents a column of data, as with the simulated data in Table 3-1.

*Table 3-1. Simulated data illustrating the relationship in our causal diagram*

| Date | Temperature | IcedCoffeeSales | $\beta$ * Temperature | $\varepsilon$ = IcedCoffeeSales − $\beta$ * Temperature |
|---|---|---|---|---|
| 6/1/2019 | 71 | $70,945 | $71,000 | $55 |
| 6/2/2019 | 57 | $56,969 | $57,000 | $31 |
| 6/3/2019 | 79 | $78,651 | $79,000 | -$349 |

For people who are familiar with linear algebra notation, we can rewrite the previous equation as:

$$\begin{pmatrix} 70,945 \\ 56,969 \\ 78,651 \end{pmatrix} = 1000 * \begin{pmatrix} 71 \\ 57 \\ 79 \end{pmatrix} + \begin{pmatrix} 55 \\ 31 \\ -349 \end{pmatrix}$$

## Causal Diagrams and Error Terms

The last column in Table 3-1, which shows the residual or error term in our regression, doesn't have any counterpart in the CD from Figure 3-2. For the sake of completeness, error terms are sometimes represented in CDs as empty circles (Figure 3-7). Feel free to do so if you find it helpful; I won't in the rest of the book because I think it makes CDs less readable. I'll just assume that any relationship in a CD that we would want to estimate comes with an implicit error term.

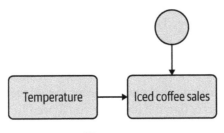

*Figure 3-7. Adding an error term to a CD*

From that perspective, CDs are all about data—variables and relationships between them. This generalizes immediately to multiple causes. Let's draw a causal diagram showing that *Temperature* and *SummerMonth* both cause *IceCreamSales* (Figure 3-8).

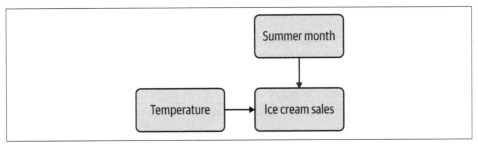

*Figure 3-8. A causal diagram with more than one cause*

Translating this CD in mathematical terms would yield the following equation:

$$IceCreamSales = \beta_T. Temperature + \beta_S. SummerMonth + \varepsilon$$

Obviously, this equation is a standard multiple linear regression, but the fact that it is based on a CD changes its interpretation. Outside of a causal framework, the only conclusion we would be able to draw from it is "an increase of one degree of temperature is associated with an increase of $\beta_T$ dollars in ice cream sales." Because correlation is not causation, it would be illegitimate to infer anything further. However, when a regression is backed by a CD, as is the case here, we can make a significantly stronger statement—namely, "unless this CD is wrong, an increase of one degree of temperature will cause an increase of $\beta_T$ dollars in ice cream sales," which is what the business cares about.

If you have a quantitative background such as data science, you may be tempted to focus on the connection between CDs and data at the expense of the connection with behaviors. It is certainly a viable path, and it has given birth to an entire category of statistical models called probabilistic graphical models. For instance, algorithms have been and are still developed to identify causal relationships in data without relying on human expertise or judgment. However, this field is still in its infancy, and when applied to real-life data, these algorithms are often unable to select between several possible CDs that lead to vastly different business implications. Business and common sense can frequently do a better job of selecting the most reasonable one. Therefore I strongly believe that you are better off using the mixed approach shown in this book's framework and accepting the idea that you'll need to use your judgment. The back and forth that CDs enable between your intuitions and your data is—literally, in many cases—where the money is.

# Fundamental Structures of Causal Diagrams

Causal diagrams can take a bewildering variety of shapes. Fortunately, researchers have been working on causality for a while now, and they have brought some order to it:

- There exist only three fundamental structures—chains, forks, and colliders—and all causal diagrams can be represented as combinations of them.
- By looking at CDs as if they were family trees, we can easily describe relationships between variables that are far away from each other in the diagram, for example by saying that one is the "descendant" or the "child" of another.

And really, that's all there is to it! We'll now see these fundamental structures in more detail, and once you have familiarized yourself with them and how to name relationships between variables, you'll be able to fully describe any CD you work with.

## Chains

A chain is a causal diagram with three boxes, representing three variables, and two arrows connecting these boxes in a straight line. To show you one, I'll have to introduce a new treat in our C-Mart example—namely, the mighty donut. For the sake of simplicity, let's assume that only one of the variables we have already seen affects the sales of donuts: *IcedCoffeeSales*. Then *Temperature*, *IcedCoffeeSales*, and *DonutSales* are causally connected (Figure 3-9).

---

*Figure 3-9. Causal diagram of a chain*

What makes this CD a chain is that the two arrows are going "in the same direction," i.e., the first arrow goes from one box to another, and the second arrow goes from that second box to the last one. This CD is an expansion of the one in Figure 3-3. It represents the fact that temperature causes sales of iced coffee, which in turn causes sales of donuts.

Let's define a few terms that will allow us to characterize the relationships between variables. In this diagram, *Temperature* is called the *parent* of *IcedCoffeeSales*, and *IcedCoffeeSales* is a *child* of *Temperature*. But *IcedCoffeeSales* is also a parent of *DonutSales*, which is its child. When a variable has a parent/child relationship with another variable we call that a *direct relationship*. When there are intermediary variables between them, we call that an *indirect relationship*. The actual count of variables that makes a relationship indirect is not generally important, so you don't have to count the number of boxes to describe the fundamental structure of the relationship between them.

In addition, we say that a variable is the *ancestor* of another variable if the first variable is the parent of another, which may be the parent of another, and so on, ending up with our second variable as a child. In our example, *Temperature* is an ancestor of *DonutSales* because it's a parent of *IcedCoffeeSales*, which is itself a parent of *DonutSales*. Very logically, this makes *DonutSales* a *descendant* of *Temperature* (Figure 3-10).

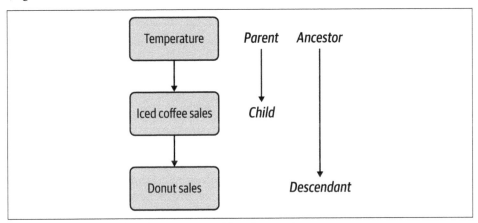

*Figure 3-10. Relationships between variables along a chain*

In this situation, *IcedCoffeeSales* is also the *mediator* of the relationship between *Temperature* and *DonutSales*. We'll explore mediation in more depth in Chapter 12. For now, let's just note that if a mediator value does not change, then the variables earlier in a chain won't influence the variables further along the chain unless they are otherwise connected. In our example, if C-Mart were to experience a shortage of iced coffee, we would expect that for the duration of that shortage, changes in temperature would not have any effect on the sales of donuts.

### Collapsing chains

The preceding causal diagram translates into the following regression equations:

$$IcedCoffeeSales = \beta_T.Temperature$$

$$DonutSales = \beta_I.IcedCoffeeSales$$

We can replace *IcedCoffeeSales* by its expression in the second equation:

$$DonutSales = \beta_I.(\beta_T Temperature) = (\beta_I\beta_T)Temperature$$

But $\beta_I\beta_T$ is just the product of two constant coefficients, so we can treat it as a new coefficient in itself: $DonutSales = \tilde{\beta}_T.Temperature$. We have managed to express *DonutSales* as a linear function of *Temperature*, which can in turn be translated into a causal diagram (Figure 3-11).

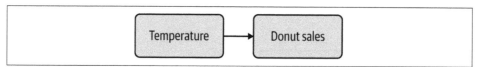

*Figure 3-11. Collapsing a CD into another CD*

Here, we have *collapsed a chain*, that is, we have removed the variable in the middle and replaced it with an arrow going from the first variable to the last. By doing so, we have effectively simplified our original causal diagram to focus on the relationship that we're interested in. This can be useful when the last variable in a chain is a business metric we're interested in and the first one is actionable. In some circumstances, for example, we might be interested in the intermediary relations between *Temperature* and *IcedCoffeeSales* and between *IcedCoffeeSales* and *DonutSales* to manage pricing or promotions. In other circumstances, we might be interested only in the relation between *Temperature* and *DonutSales*—for example, to plan for inventory.

The transitivity property of linear algebra also applies here: if *DonutSales* caused another variable, then that chain could also be collapsed around *DonutSales*, and so on.

## Expanding chains

The collapsing operation can obviously be reversed: we can go from our last CD to the previous one by adding the *IcedCoffeeSales* variable in the middle. More generally, we say that we are *expanding a chain* whenever we inject an intermediary variable between two variables currently connected by an arrow. For example, let's say that we start with the relationship between *Temperature* and *DonutSales* (Figure 3-11). This causal relationship translates into the equation *DonutSales* = $\tilde{\beta}_T$*Temperature*. Let's assume that *Temperature* affects *DonutSales* only through *IcedCoffeeSales*. We can add this variable in our CD, which brings us back to our original CD from Figure 3-8 (Figure 3-12).

*Figure 3-12. Expanding a CD into another CD*

Expanding chains can be useful to better understand what's happening in a given situation. For example, let's say that temperature increased but sales of donuts did not. There could be two potential reasons for that:

- First, the increase in temperature did not increase the sales of iced coffee, perhaps because the store manager has been more aggressive with the AC. In other words, the first arrow in Figure 3-11 disappeared or weakened.

- Alternatively, the increase in temperature *did* increase the sales of iced coffee, but the increase in the sales of iced coffee did not increase the sales of donuts, e.g., because people are buying the newly offered cookies instead. In other words, in Figure 3-11, the first arrow is unchanged but the second one disappeared or weakened.

Depending on which one is true, you might take very different corrective actions—either turning off the AC or changing the price of cookies. In many cases, looking at the variable in the middle of a chain, namely the mediator, will allow you to make better decisions.

Because chains can be collapsed or expanded at will, in general we do not explicitly indicate when it has been done. It's always assumed that any arrow could potentially be expanded to highlight an intermediary variable along the way.

This also implies that the definition of "direct" and "indirect" relationships mentioned earlier relates to a specific representation of a CD: when you collapse a chain, two variables that had an indirect relationship now have a direct relationship.

# Forks

When a variable causes two or more effects, the relationship creates a *fork*. *Temperature* causes both *IcedCoffeeSales* and *IceCreamSales*, so a representation of this fork is shown in Figure 3-13.

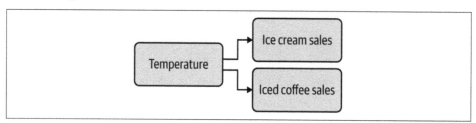

*Figure 3-13. A fork between three variables*

This CD shows that *Temperature* influences both *IceCreamSales* and *IcedCoffeeSales*, but that they do not have a causal relationship with each other. If it is hot out, demand for both ice cream and iced coffee increases, but buying one does not make you want to buy the other, nor does it make you less likely to buy the other.

This situation where two variables have a common cause is very frequent but also potentially problematic, because it creates a correlation among these two variables. It makes sense that when it is hot out, we will see an increase in sales of both, and when it is cold fewer people will want both. A linear regression predicting *IceCreamSales* from *IcedCoffeeSales* would be fairly predictive, but here correlation does not equal causation, and since we know that the causal impact is 0, the coefficient provided by the model would not be accurate.

Another way to look at this relationship is that if C-Mart experienced a shortage of iced coffee, we would not expect to see a change in the sales of ice cream. More generally, it would be only a slight exaggeration to say that forks are one of the main roots of evil in the world of data analysis. Whenever we observe a correlation between two variables that doesn't reflect direct causality between them (i.e., neither is the cause of the other), more often than not it will be because they share a common cause. From that perspective, one of the main benefits of using CDs is that they can show very clearly and intuitively what's going on in those cases and how to correct for it.

Forks are also typical of situations where we look at demographic variables: age, gender, and place of residence all cause a variety of other variables that may or may not cause each other. You can picture a demographic variable such as age as being at the root of a fork with many teeth.

A question that sometimes comes up when you have a fork in the middle of a CD is whether you can still collapse the chain around it. For example, let's say that we're interested in analyzing the relationship between *SummerMonth* and *IcedCoffeeSales* using the CD in Figure 3-14.

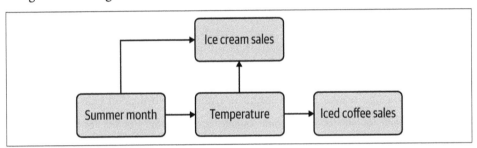

*Figure 3-14. A CD with a fork and a chain*

In this CD, there's a fork between *SummerMonth* on one side and *IceCreamSales* and *Temperature* on the other, but there's also a chain *SummerMonth* → *Temperature* → *IcedCoffeeSales*. Can we collapse the chain?

In this case, yes. We'll see in Chapter 5 how to determine when a variable is a confounder of a relationship; here it will suffice to say that *IceCreamSales* is not a confounder of the relationship between *SummerMonth* and *IcedCoffeeSales*, which is the one we're interested in. Therefore we can simplify our CD (Figure 3-15).

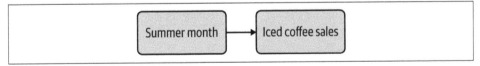

*Figure 3-15. A collapsed version of the previous CD*

Similarly, if we were interested in the relationship between *SummerMonth* and *IceCreamSales* in Figure 3-14, we could neglect *IcedCoffeeSales* but not *Temperature*.

Because forks are so important to causal analysis, we'll sometimes want to represent them even when we don't know the joint cause. When that's the case, we'll represent the unknown fork with a two-headed arrow (Figure 3-16).

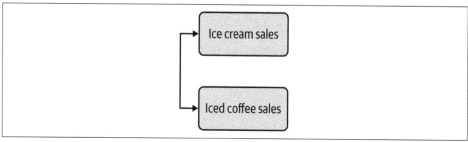

*Figure 3-16. A fork with an unknown joint cause*

The two-headed arrow also looks like the two variables are causing each other. This is by design, and we'll also use a two-headed arrow when we observe a correlation between two variables but we don't know which is causing which. Thus, a two-headed arrow encompasses the three possible reasons why two variables A and B would appear correlated: A causes B, B causes A, and/or A and B share a cause. Sometimes we'll use a two-headed arrow as a placeholder until we clarify the true reason; if we don't care about the reason, we may simply retain the two-headed arrow in our final CD.

## Colliders

Very few things in the world have only one cause. When two or more variables cause the same outcome, the relationship creates a *collider*. Since C-Mart's concession stand sells only two flavors of ice cream, chocolate and vanilla, a causal diagram representing taste and ice cream purchasing behavior would show that appetite for either flavor would cause past purchases of ice cream at the stand (Figure 3-17).

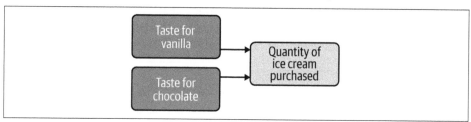

*Figure 3-17. CD of a collider*

Colliders are a common occurrence, and they can also be an issue in data analysis. A collider is in a sense the opposite of a fork, and the problems with them are also symmetric: a fork is problematic if we *don't control* for the joint cause whereas a collider is a problem if we *do control* for the joint effect. We'll explore these issues further in Chapter 5.

To recap this section, chains, forks, and colliders represent the only three possible ways for three variables to be related to each other in a CD. They are not exclusive of each other, however, and it's actually reasonably common to have three variables that exhibit all three structures at the same time, as was the case in our very first example (Figure 3-18).

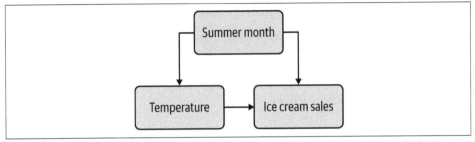

*Figure 3-18. A three-variable CD containing a chain, a fork, and a collider at the same time*

Here, *SummerMonth* influences *IceCreamSales* as well as *Temperature*, which itself influences *IceCreamSales*. The causal relationships at play are reasonably simple and easy to grasp, but this graph also contains all three types of basic relationships:

- A chain: *SummerMonth → Temperature → IceCreamSales*
- A fork, with SummerMonth causing both *Temperature* and *IceCreamSales*
- A collider, with *IceCreamSales* being caused by both *Temperature* and *SummerMonth*

Another thing to note in a situation like this one is that variables have more than one relationship with each other. For example, *SummerMonth* is the parent of *IceCreamSales* because there is an arrow going directly from the former to the latter (a direct relationship); but at the same time, *SummerMonth* is also indirectly an ancestor of *IceCreamSales* through the chain *SummerMonth → Temperature → IceCreamSales* (an indirect relationship). So you can see these are not exclusive!

While a CD is always made of these three structures, it is not static. A CD can be transformed by modifying the variables themselves as well as their relationships, as we'll now see.

# Common Transformations of Causal Diagrams

Chains, forks, and colliders take the variables in a CD as given. But in the same way that a chain can be collapsed or expanded, variables can themselves be sliced or aggregated to "zoom" in and out of specific behaviors and categories. We may also decide to modify the arrows—for example, when we're faced with otherwise intractable cycles.

## Slicing/Disaggregating Variables

Forks and colliders are often created when you slice or disaggregate a variable to reveal its components. In a previous example, we looked at the relationship between *Temperature* and *DonutSales*, where *IcedCoffeeSales* was the mediator (Figure 3-19).

*Figure 3-19. The chain that we will slice*

But maybe we want to split *IcedCoffeeSales* by type to better understand demand dynamics. This is what I mean by "slicing" a variable. This is allowed per the rules of linear algebra, because we can express the total iced coffee sales as the sum of sales by type, say Americano and latte:

$$IcedCoffeeSales = IcedAmericanoSales + IcedLatteSales$$

Our CD would now become Figure 3-20, with a fork on the left and a collider on the right.

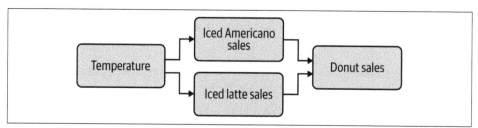

*Figure 3-20. A chain where the mediator has been sliced*

Each slice of the variable would now have its own equation:

$$IcedAmericanoSales = \beta_{TA}.Temperature$$

$$IcedLatteSales = \beta_{TL}.Temperature$$

Since the effect of *Temperature* is completely mediated by our *IcedCoffeeSales* slices, we can create a unified multiple regression for *DonutSales* as follows:

$$DonutSales = \beta_{IA}.IcedAmericanoSales + \beta_{IL}.IcedLatteSales$$

This would allow you to understand more finely what's happening—should you plan for the same increase in sales in both types when temperature increases? Do they both have the same effect on *DonutSales* or should you try to favor one of them?

## Aggregating Variables

As you may have guessed, slicing variables can be reversed, and more generally we can aggregate variables that have the same causes and effects. This can be used to aggregate and disaggregate data analysis by product, region, line of business, and so on. But it can also be used more loosely to represent important causal factors that are not precisely defined. For example, let's say that *Age* and *Gender* both impact *Taste-ForVanilla* as well as the propensity to buy ice cream at C-Mart concession stands, *PurchasedIceCream* (Figure 3-21).

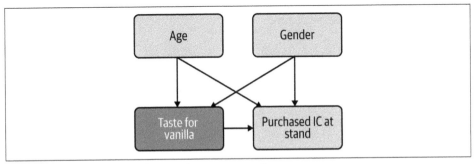

*Figure 3-21. Age and Gender are shown separately*

Because *Age* and *Gender* have the same causal relationships, they can be aggregated into a *DemographicCharacteristics* variable (Figure 3-22).

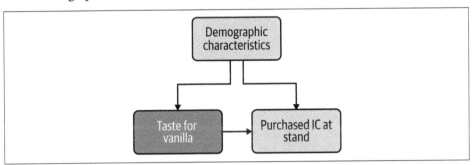

*Figure 3-22. CD where Age and Gender are aggregated into a single variable*

In this case, we obviously don't have a single column in our data called "Demographic Characteristics" or "Demographics"; we're simply using that variable in our CD as a shortcut for a variety of variables that we may or may not want to explore in further detail later on. Let's say that we want to run an A/B test and understand the causal relationships at hand. As we'll see later, randomization can allow us to control for demographic factors so that we won't have to include them in our analysis, but we might want to include them in our CD of the situation without randomization. If need be, we can always expand our diagram to accurately represent the demographic variables involved. Remember, however, that *any variable can be split, but only variables that have the same direct and indirect relationships can be aggregated.*

---

## Technical Deeper Dive: Aggregating Variables and Linear Algebra

Using a summary *Demographics* variable may look like a dubious sleight of hand, but it is actually completely sound from the perspective of linear algebra. If you'll take my word for it or your linear algebra is too rusty, you can just skip this box and not worry about it. But if you want to check the math, here it is.

We can treat our variables as vectors:

$$TasteForVanilla = \begin{pmatrix} 3 \\ 15 \\ 8 \\ \vdots \\ 17 \end{pmatrix}, Age = \begin{pmatrix} 23 \\ 78 \\ 52 \\ \vdots \\ 41 \end{pmatrix} \text{ and } Gender = \begin{pmatrix} 0 \\ 1 \\ 1 \\ \vdots \\ 0 \end{pmatrix}$$

We now have the corresponding equation:

$$TasteForVanilla = \beta_A.Age + \beta_G.Gender$$

But we can also attach our vectors together in a matrix (which is technically how linear regression models are traditionally solved):

$$Demographics = (Age \quad Gender) = \begin{pmatrix} 23 & 0 \\ 78 & 1 \\ 52 & 1 \\ \vdots & \vdots \\ 41 & 0 \end{pmatrix}$$

This allows us to rewrite the previous equation as:

$$TasteForVanilla = (Age \quad Gender) * \begin{pmatrix} \beta_a \\ \beta_g \end{pmatrix} = Demographics * \vec{\beta} \text{ with } \vec{\beta} = \begin{pmatrix} \beta_a \\ \beta_g \end{pmatrix}$$

---

We have now expressed *TasteForVanilla* as a linear function of *Demographics* in vector notation. In other words, as long as we're only aggregating variables that have exactly the same relationships with other variables, we're on solid mathematical grounds.

## What About Cycles?

In the three fundamental structures that we've seen, there has been only one arrow between two given boxes. More generally, it was not possible to reach the same variable twice by following the direction of arrows (e.g., A → B → C → A). A variable could be the effect of one variable and the cause of another, but it could not be at the same time the cause and the effect of one variable.

In real life, however, we often see variables that influence each other causally. This type of CD is called a *cycle*. Cycles can arise for a variety of reasons; two of the most common in behavioral data analysis are substitution effects and feedback loops. Fortunately, there are some workarounds that will allow you to deal with cycles when you encounter them.

### Understanding cycles: Substitution effects and feedback loops

Substitution effects are a cornerstone of economics theory: customers might *substitute* a product for another, depending on the products' availability and price and the customers' desire for variety. For example, customers coming to the C-Mart concession store might choose between iced coffee and hot coffee based not only on temperature but also on special promotions and how often they had coffee this week. Therefore, there is a causal relationship from purchases of iced coffee to purchases of hot coffee, and another causal relationship in the opposite direction (Figure 3-23).

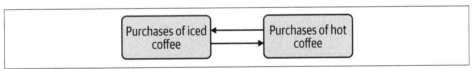

*Figure 3-23. A CD with a substitution effect generating a cycle*

One thing to note is that the direction of the arrows shows the direction of causality (what is the cause and what is the effect), not the sign of the effect. In all of the CDs we looked at previously, the variables had a positive relationship where an increase in one caused an increase in the other. In this case, the relationships are negative, where an increase in one variable will cause a decrease in the other. The sign of the effect does not matter for causal diagrams, and a regression will be able to sort out the sign for the coefficient correctly as long as you correctly identify the relevant causal relationships.

Another common cycle is a feedback loop, where a person modifies their behavior in reaction to changes in the environment. For example, a store manager at C-Mart might keep an eye on the length of waiting lines and open new lines if the existing ones get too long, so that customers don't give up and just leave (Figure 3-24).

*Figure 3-24. Example of a feedback loop generating a cycle*

## Managing cycles

Cycles reflect situations that are often complex to study and manage, which is why a whole field of research, called *systems thinking*, has sprouted for that purpose.[2] Complex mathematical methods, such as structural equation modeling, have been developed to deal accurately with cycles, but their analysis would take us beyond the scope of this book. I would be remiss, however, if I didn't give you any solution, so I'll mention two rules of thumb that should keep you from getting stuck with cycles.

The first one is to pay close attention to timing. In almost all cases, it takes some time for one variable to influence another, which means you can "break the cycle" and turn it into an "acyclical" CD, i.e., a CD without cycles (which you can then analyze with the tools presented in this book), by looking at your data at a more granular level of time. For example, let's say that it takes 15 minutes for a store manager to react to an increasing waiting time by getting new lines open, and it similarly takes 15 minutes for customers to adjust their perception of waiting time. In that case, by clarifying the temporal order of things, we can split the waiting time variable in our CD (Figure 3-25).

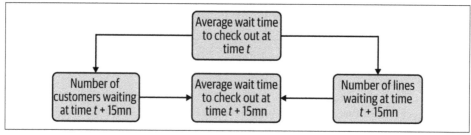

*Figure 3-25. Breaking a feedback loop into time increments*

2 Interested readers are referred to *Thinking in Systems: A Primer* by Meadows and Wright (2008), as well as *The Fifth Discipline: The Art & Practice of the Learning Organization* by Senge (2010).

I'll explain this CD one piece at a time. On the left, we have an arrow from average waiting time to number of customers waiting:

$$NbCustomersWaiting(t + 15mn) = \beta_1 . AvgWaitingTime(t)$$

This means that the number of customers waiting at, say, 9:15 a.m. would be expressed as a function of the average waiting time at 9:00 a.m. Then the number of customers waiting at 9:30 a.m. would have the same relation to the average waiting time at 9:15 a.m. and so on.

Similarly, on the right, we have an arrow from average waiting time to number of lines open:

$$NbLinesOpen(t + 15mn) = \beta_2 . AvgWaitingTime(t)$$

This means that the number of lines open at 9:15 a.m. would be expressed as a function of the average waiting time at 9:00 a.m. Then the number of lines open at 9:30 a.m. would have the same relation to the average waiting time at 9:15 a.m. and so on.

Then in the middle, we have causal arrows from the number of customers waiting and from the number of lines open to the average waiting time. Assuming linear relationships here for the sake of simplicity, this would translate into the following equation:

$$AvgWaitingTime(t) = \beta_3 . NbCustomersWaiting(t) + \beta_4 . NbLinesOpen(t)$$

In reality, the assumption of linear relationships is *very* unlikely to be true in this case. Specific models have been developed for queues, or for time-to-event variables (e.g., survival analysis). These models are part of the broader category of GLMs, and as such, a good rule of thumb is that for our purposes they'll behave like logistic regressions.

This means that the average waiting time for customers reaching the checkout lines at 9:15 a.m. depends on the number of customers already present and the number of checkout lines open at 9:15 a.m. Then the average waiting time for customers reaching the checkout lines at 9:30 a.m. depends on the number of customers already present and the number of checkout lines open at 9:30 a.m. and so on.

By breaking down variables into time increments, we have been able to create a CD where there is no cycle in the strict sense. We can estimate the three preceding linear regression equations without introducing any circular logic.

The second rule of thumb for dealing with cycles is to simplify your CD and to keep only the arrows along the causal path you are most interested in. Feedback effects (where a variable influences the variable that just influenced it) are generally smaller, and often much smaller, than the first effect and can be ignored as a first approximation.

In our example of iced and hot coffee, you might be worried that the increase in sales of iced coffee when it is hot will decrease the sale of hot coffee; this is a reasonable concern that you should investigate. However, it's unlikely that the decrease in sales of hot coffee would in turn trigger a further increase in sales of iced coffee, and you can ignore that feedback effect in your CD (Figure 3-26).

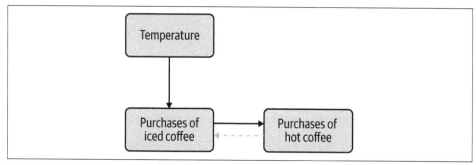

*Figure 3-26. Simplifying a CD by neglecting certain relationships*

In Figure 3-26, we delete the arrow from purchases of hot coffee to purchases of iced coffee and ignore that relationship as a reasonable approximation.

Once again, this is just a rule of thumb, and certainly not a blanket invitation to disregard cycles and feedback effects. These should be represented fully in your complete CD to guide future analyses.

## Paths

Having seen the various ways variables can interact, we can now introduce one last concept that encompasses all of them: *paths*. We say that there is a path between two variables *if there are arrows between them, regardless of the direction of the arrows, and if no variable appears twice along the way*. Let's see what that looks like in a CD we have seen before (Figure 3-27).

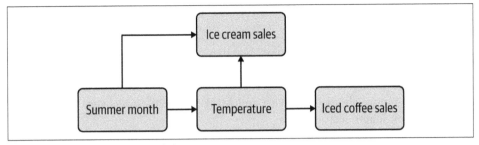

*Figure 3-27. Paths in a causal diagram*

In the previous CD, there are two paths from *SummerMonth* to *IcedCoffeeSales*:

- One path along the chain *SummerMonth → Temperature → IcedCoffeeSales*
- A second path through *IceCreamSales, SummerMonth → IceCreamSales ← Temperature → IcedCoffeeSales*

This means that a chain is a path, but so are a fork or a collider! Also note that two different paths between two variables can also share some arrows, as long as there is at least one difference between them, as is the case here: the arrow from *Temperature* to *IcedCoffeeSales* appears in both paths.

However, the following is not a valid path between *Temperature* and *IcedCoffeeSales* because *Temperature* appears twice:

- *Temperature ← SummerMonth → IceCreamSales ← Temperature → IcedCoffeeSales*

One consequence of these definitions is that if you pick two different variables in a CD, there is always at least one path between them. The definition of paths may seem so broad that it is useless, but as we'll see in Chapter 5, paths will actually play a crucial role in identifying confounders in a CD.

# Conclusion

Correlation is not causation because confounders can introduce bias in our analyses. Unfortunately, as we've seen through examples, simply throwing all available variables and the kitchen sink into a regression is not sufficient to resolve confounding. Worse, controlling on the wrong variables can introduce spurious correlations and create new biases.

As a first step toward unbiased regression, I introduced a tool known as causal diagrams. CDs may be the best analytical tool you've never heard of. They can be used to represent our intuitions about causal relationships in the real world, as well as causal relationships between variables in our data; but they are most powerful as a bridge between the two, allowing us to connect our intuition and expert knowledge to observed data, and vice versa.

CDs can get convoluted and complex, but they are based on three simple building blocks: chains, forks, and colliders. They can also be collapsed or expanded, sliced, or aggregated, according to simple rules that are consistent with linear algebra. If you want to know more about CDs, Pearl and Mackenzie (2018) is a very readable and enjoyable book-length introduction.

The full power of CDs will become apparent in Chapter 5, where we'll see that they allow us to optimally handle confounders in regression, even with nonexperimental data. But CDs are also helpful more broadly, to help us think better about data. In the next chapter, as we get into cleaning and prepping data for analysis, they will allow us to mitigate biases in our data prior to any analysis. This will give you the opportunity to get more familiar with CDs in a simple setting.

# CHAPTER 4

# Building Causal Diagrams from Scratch

*At this point, you may be wondering where the [causal diagram] comes from. It's an excellent question. It may be the question. A [CD] is supposed to be a theoretical representation of the state-of-the-art knowledge about the phenomena you're studying. It's what an expert would say is the thing itself, and that expertise comes from a variety of sources. Examples include economic theory, other scientific models, conversations with experts, your own observations and experiences, literature reviews, as well as your own intuition and hypotheses.*

—Scott Cunningham, *Causal Inference: The Mixtape* (2021)

Our goal in this book is always to measure the impact of one variable on another, which we can represent as a "starter" CD (Figure 4-1).

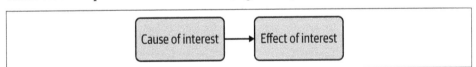

*Figure 4-1. The simplest possible CD*

Once you have drawn that relationship, what comes next? How can you know what other variables you should include or not? Many authors say you should rely on expert knowledge, which is fine if you work in an established field like economics or epidemiology. But my perspective in this book is that you're likely "behavioral scientist number one" in your organization and therefore you need to be able to start from a blank slate.

In this chapter, I'll outline a recipe to get you from the basic CD in Figure 4-1 to a workable CD. As we go through the process, please keep in mind our ultimate goal of understanding what drives behaviors so that we can draw relevant and actionable conclusions for our business. Our objective is not to build a complete and precise knowledge of the whole world. Shortcuts and approximations are fair game, and everything should be assessed according to a single criterion: is this helping me with my business goal?

In addition, the recipe I'll outline is not a mechanical algorithm that you could follow blindly to get to the right CD. On the contrary, business sense, common sense, and data insights will be crucial. We'll go back and forth between our qualitative understanding of the causal situation at hand and the quantitative relationships present in the data, cross-checking one with the other until we feel that we have a satisfactory result. "Satisfactory" is an important word here: in applied settings, you usually can't tell your manager that you'll give them the right answer in three years. You need to give them the least bad answer possible in the short term, while planning the data collection work that will improve your answer over the years.

In the next section, I'll present the business problem for this chapter and the corresponding variables of interest. We'll then progressively build out the corresponding CD along the following recipe, with one section per step:

- Identify variables that could/should potentially be included in the CD.
- Determine if the variables should be included.
- Iterate the process as needed.
- Simplify the diagram.

Let's get started!

# Business Problem and Data Setup

For this section, we'll be working with a real-world data set of hotel bookings for two hotels located within the same city.[1] The data and packages that we'll use are described in the next subsection, and we'll then do a deeper dive to understand the relationship of interest.

## Data and Packages

The GitHub folder for this chapter (*https://oreil.ly/BehavioralDataAnalysisCh4*) contains the CSV file *chap4-hotel_booking_case_study.csv* with the variable listed in Table 4-1.

---

1 Nuno Antonio, Ana de Almeida, and Luis Nunes, "Hotel booking demand data sets," *Data in Brief*, 2019. *https://doi.org/10.1016/j.dib.2018.11.126*

*Table 4-1. Variables in data file*

| Variable name | Variable description |
| --- | --- |
| *NRDeposit* (NRD) | Binary 0/1, whether the reservation had a nonrefundable deposit |
| *IsCanceled* | Binary 0/1, whether the reservation was canceled or not |
| *DistributionChannel* | Categorical variable with values "Direct," "Corporate," "TA/TO" (travel agent/travel organization), "Other" |
| *CustomerType* | Categorical variable with values "Transient," "Transient-Party," "Contract," "Group" |
| *MarketSegment* | Categorical variable with values "Direct," "Corporate," "Online TA," "Offline TA/TO," "Groups," "Other" |
| *Children* | Integer, number of children in the reservation |
| *ADR* | Numeric, average daily rate, total reservation amount/number of days |
| *PreviousCancellation* | Binary 0/1, whether the customer previously canceled a reservation or not |
| *IsRepeatedGuest* | Binary 0/1, whether the customer has previously made a reservation at the hotel |
| *Country* | Categorical, country of origin of the customer |
| *Quarter* | Categorical, quarter of year for the reservation |
| *Year* | Integer, year of the reservation |

In this chapter, we'll use the following packages in addition to the standard ones called out in the preface:

```R
## R
library(rcompanion) # For Cramer V correlation coefficient function
library(car) # For VIF diagnostic function

## Python
from math import sqrt # For Cramer V calculation
from scipy.stats import chi2_contingency # For Cramer V calculation
```

## Understanding the Relationship of Interest

We'll try to answer the question, "Does the type of deposit impact the cancellation rate of bookings?" as expressed in Figure 4-2.

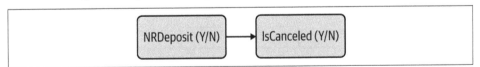

*Figure 4-2. Causal relationship of interest*

Let's start by looking at the base cancellation rate by deposit type (I like to look at both the absolute numbers and the percentages, in case some categories have very small numbers):

```
## R (output not shown)
with(dat, table(NRDeposit, IsCanceled))
with(dat, prop.table(table(NRDeposit, IsCanceled), 1))

## Python
table_cnt = dat_df.groupby(['NRDeposit', 'IsCanceled']).\
agg(cnt = ('Country', lambda x: len(x)))
print(table_cnt)

table_pct = table_cnt.groupby(level=0).apply(lambda x: 100 * x/float(x.sum()))
print(table_pct)
                         cnt
NRDeposit IsCanceled
0         0            63316
          1            23042
1         0               55
          1              982
                         cnt
NRDeposit IsCanceled
0         0        73.318048
          1        26.681952
1         0         5.303761
          1        94.696239
```

We can see that the overwhelming majority of bookings have no deposit, with a cancellation rate of about 27%. On the other hand, bookings with nonrefundable deposits (*NRDeposit*) have a very high cancellation rate. On the face of it, this correlation is surprising. Would changing our policy to "no deposit" for everyone reduce cancellation rates? Behavioral common sense tells us that more likely, the hotels are requesting nonrefundable deposits for "high risk" bookings and there is a confounder, as represented in Figure 4-3.

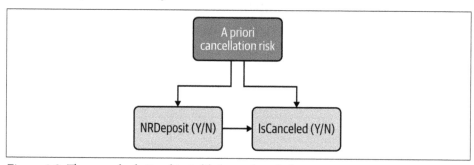

*Figure 4-3. The causal relationship is likely confounded*

We went pretty quickly from Figure 4-2 to Figure 4-3, but it was an important step: the CD in Figure 4-2 represents a basic business analytics question, "What is the causal relationship between deposit type and cancellation rate?" On the other hand, the CD in Figure 4-3 represents a more informed behavioral hypothesis: "A nonrefundable deposit seems to increase cancellation rate, but that relationship is probably confounded by factors we'll need to determine."

A nice aspect of using CDs for behavioral data analysis is that they are a great collaboration tool. Anyone in your organization with minimal knowledge of CDs can look at Figure 4-3 and say, "Well yeah, we require nonrefundable deposits for holiday bookings and these often get canceled because of weather," or any other tidbit of behavioral knowledge that you couldn't get otherwise.

At this point, the best next step would be a randomized experiment: assign refundable or nonrefundable deposits to a random sample of customers and you'll be able to confirm or disprove your behavioral hypothesis. However, you may not be able to do so, or not yet. In the meantime, we'll try to deconfound the relationship by identifying relevant variables to include.

# Identify Candidate Variables to Include

When trying to identify potential variables to include, a natural inclination is to start with the data you have available. This inclination is misleading, akin to the drunk person who looks for their house keys not where they lost them, but under a streetlamp because there is more light there. By doing so, you may be ignoring the most important variables simply because they're not under your nose. You're also more likely to take the variables in your data at face value, and not question whether they are the best representation of what's happening in the real world.

For instance, categorical variables in your data are likely to represent a business-centric perspective rather than a customer-centric perspective, and it may be more appropriate to aggregate some categories together, or even to merge different variables into new ones. In our case, we have a variable *MarketSegment* and one for the number of children in the reservation. We can confirm by looking at the data that very few corporate customers bring children with them. Therefore we could consider creating a new categorical variable with categories "corporate without children," "noncorporate without children," and "noncorporate with children," setting aside the corporate customers with children as outliers worthy of a separate investigation (maybe the seed for tailored services?).

Instead of falling for this "What You See Is All There Is" bias,[2] we'll start with the behavioral categories outlined in Chapter 2, from the action backward:

---

2 This colorful label was popularized by behavioral scientist Daniel Kahneman.

- Actions
- Intentions
- Cognition and emotions
- Personal characteristics
- Business behaviors

Finally, variables in each one of these categories may be affected by time trends such as linear trends or seasonality, so we'll add these by default at the end of this section. In order to reinforce the focus on qualitative intuitions, we won't look at any data until the next section, about validating relationships. Replacing our a priori cancellation risk (as well as other potential confounders) by these categories, our CD now looks like Figure 4-4, with a bunch of unobserved variables added to our two variables of interest.

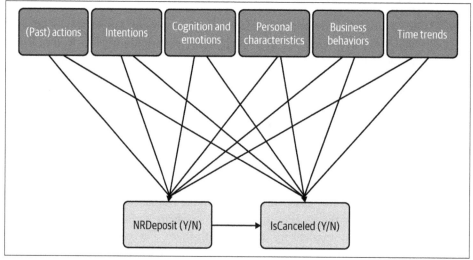

*Figure 4-4. Updated CD with categories of potential variables to include*

For each of these categories, we'll now look for variables that may be a cause of either of our two variables of interest.

# Actions

When looking for variables to include in the actions category, we're usually trying to identify past behaviors from the customer that may affect whether the hotel required a nonrefundable deposit (NRD).

An obvious candidate in this case is whether the customer previously canceled. Maybe the hotel is more likely to request an NRD from customers who flaked in the past. It is also conceivable that whatever caused them to cancel in the past is also more likely to make them cancel in the future.

More generally, when one of our variables of interest is itself an action, past behavior is often a good predictive variable to include, even just as a proxy for unobserved personal characteristics. We have two variables related to past behavior in our data: *PreviousCancellation* and *IsRepeatedGuest*. Figure 4-5 shows our updated CD, with the unchanged portions in gray.

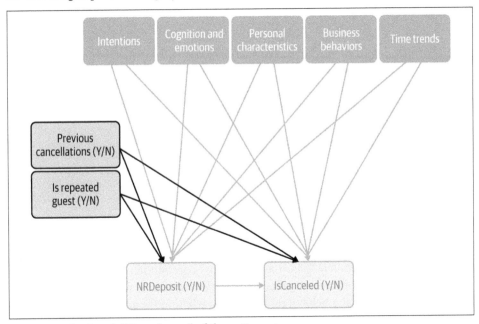

*Figure 4-5. Updated CD at the end of the actions step*

This is not to say that these are the only relevant past behaviors; they just happen to be the only ones I thought of and that we had data for. Hopefully, you can think of others!

# Intentions

Intentions are easy to overlook in data analysis, because they are often missing from our existing data. However, they are one of the most important drivers of behaviors, and they can often be revealed by interviewing customers and employees. They thus represent one of the best illustrations of the benefits of not just looking at existing, available data, but adopting a "behaviors first" approach.

In this case, there are two intentions I can think of: the reason for the trip and the reason for the cancellation (Figure 4-6).

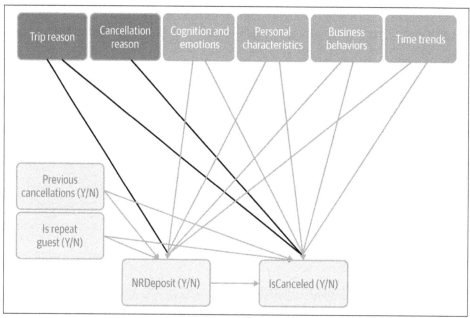

*Figure 4-6. Adding intentions to our CD*

Note that I represented *TripReason* as a potential confounder, i.e., with an arrow to both of our variables of interest, whereas *CancellationReason* affects only *IsCanceled*. At this point, this is just a behavioral hunch, namely that the reason for the cancellation doesn't affect the type of deposit. My rationale is that the reason for cancellation is not known at the time of the deposit.

Figure 4-6 also shows the versatility of CDs for behavioral analysis: we can put down these two potential variables in our CD even without knowing the actual list of reasons in either case, which we'll determine later through interviews. For the time being, we can note that three of the variables available in our data appear to be affected by trip reason and can be included as such: *CustomerType*, *MarketSegment,* and *DistributionChannel* (Figure 4-7). We'll also revisit these variables in the subsection about personal characteristics.

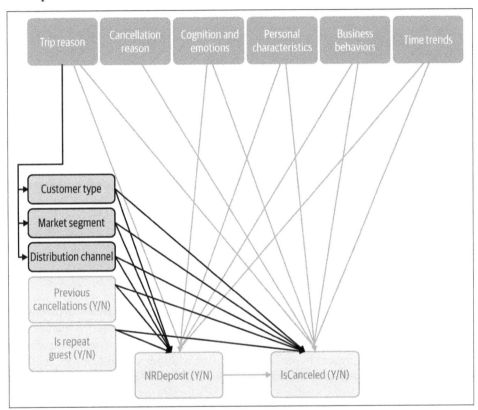

*Figure 4-7. Updated CD at the end of the intention step*

## Cognition and Emotions

When trying to identify relevant social, psychological, or cognitive phenomena for an analysis, I like to zoom in on specific decision points. Here that would be when the customer makes the reservation and when they cancel it.

At the first decision point, customers may not understand that their deposit is nonrefundable, or they may forget it. At the second decision point, they may treat their deposit as a sunk cost and not make efforts to keep their reservations (Figure 4-8).

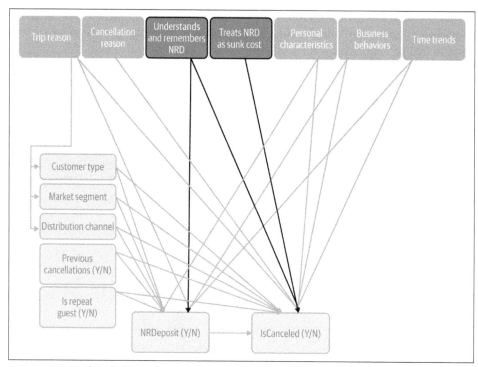

*Figure 4-8. Updated CD at the end of the cognition and emotions step*

## Personal Characteristics

As mentioned in Chapter 2, demographic variables are often valuable not so much for themselves but as proxies for other personal characteristics such as personality traits. The challenge at this step is therefore to resist the pull of whatever demographic variables are present in our data, and stick with our causal-behavioral mindset. A good way to do so is to think about traits first, before looking at demographic variables.

## Traits

Based on our knowledge of personality psychology, good candidate traits to cause our cancellation behavior are conscientiousness and neuroticism: it seems plausible that less organized and more carefree people are more likely to end up cancelling a reservation (Figure 4-9).

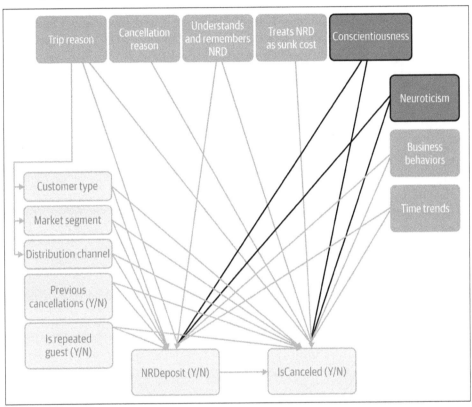

*Figure 4-9. CD updated with personality traits*

## Demographic variables

We noted earlier that we have corporate and noncorporate customers making reservations at the hotels. Beyond the reasons for the trip and for the cancellation, this also affects some other personal characteristics such as price elasticity and income, both of which would affect our two variables of interest. Let's group these under the heading "financial characteristics." They are probably somewhat captured by the three variables we saw earlier, *CustomerType, MarketSegment,* and *DistributionChannel,* as well as by a few other variables in our data such as *Children, ADR* (the average daily rate, i.e., the price for a night), and *Country* (Figure 4-10).

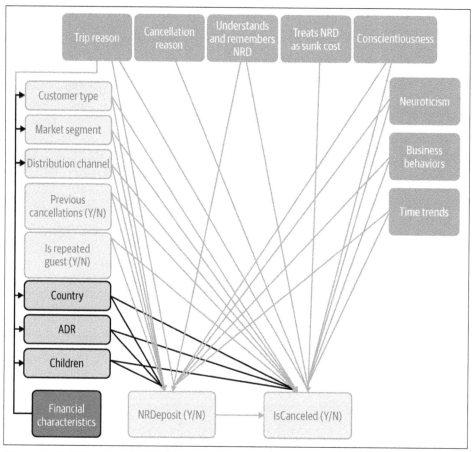

*Figure 4-10. CD updated with demographic variables*

## Business Behaviors

Business behaviors often play a big role in the relationships we're investigating, but they can easily be overlooked and tricky to integrate.

In this example, business rules obviously play an important role, as they determine which customers will have to provide an NRD. In that sense, they influence *all* the arrows in the CD going into *NRDeposit*. We can account for this influence in several ways, depending on the forms it takes.

A business rule may explicitly connect two observable variables (possibly including our variables of interest). Here for instance, we can imagine a business rule stating that all customers who have previously canceled must now provide an NRD. By listing all such rules, we can confirm or disprove all arrows from an observed variable directly into *NRDeposit*. This may also surface variables that are involved in business rules but are not yet in our data: for example, we could imagine that customers who don't provide a proof of ID at the time of the reservation must pay an NRD. I said "not yet in our data" because by definition any criterion that is part of a business rule is observable, even if it's not captured in a database.[3]

Alternatively, a business rule may be best represented as an additional intermediary variable. For instance, if all reservations during the Christmas holidays must be backed by an NRD, we could create a *ChristmasHolidays* binary variable with an arrow to *NRDeposit*. That variable would then mediate the effect of other variables such as *CustomerType* or *Children* on *NRDeposit*.

We don't know what business rules the two hotels in our example apply so we have to leave that subsection as something that we would want to explore through later interviews.

## Time Trends

Finally, there might be some global time trends in our data, such as a progressive increase in the number of reservations requiring an NRD paralleled by a progressive, but unrelated, increase in the cancellation rate. In addition, given how seasonal the hotel industry is, there are probably some cyclical aspects that we would want to capture (Figure 4-11).

---

3 Think about it: how would the rule be implemented otherwise?

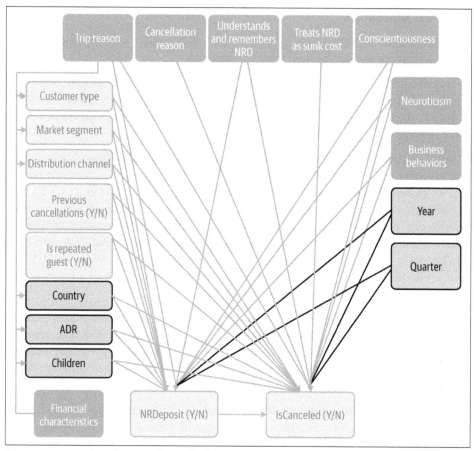

*Figure 4-11. Updated CD at the end of the time trends step*

In the present case, the *Year* and *Quarter* variables only capture trends and cycles. Sometimes it can also make sense to include binary variables to account for specific events that make a certain year stand out or mark a permanent change. An obvious example is COVID-19, which, when the dust settles, will turn out to have been a temporary blip in certain sectors but the beginning of a sea change in others.

With that final addition, Figure 4-11 now has a raft of candidate variables, some observable and some not. In the next section, we'll see how to confirm which observable variables to keep.

# Validate Observable Variables to Include Based on Data

Let's look at the observable variables we have as candidates at the end of the identification phase (Figure 4-12).

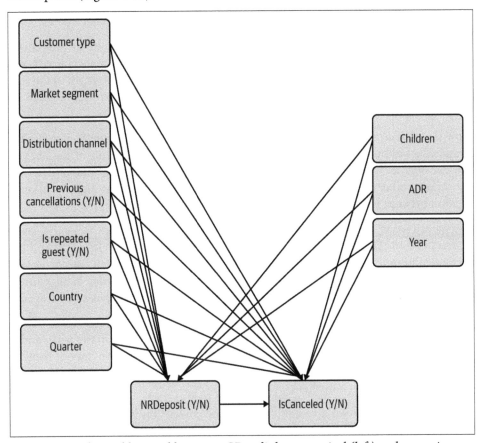

*Figure 4-12. Observable variables in our CD, split by categorical (left) and numeric (right)*

In this specific example, all of these observable variables are tentatively connected to both of our variables of interest. This is the default situation, but in some cases you may have a very strong a priori rationale to connect a predictor variable to only one of your variables of interest (that was the case, for example, with some of our unobserved variables). When in doubt, I would err on the side of caution and include both connections.

In Figure 4-12, the observable variables are split between categorical (on the left of the CD) and numeric (on the right of the CD). These two types of data call for different quantitative tools, so we'll see them in turn.

# Relationships Between Numeric Variables

Our first step will be to look at the correlation matrix for all the numeric variables in our data. A useful but dirty trick is to convert binary variables to 0/1 (if they're not already in that format) so that you can treat them as numeric. It allows you to get a sense of the correlation between variables, but don't tell your statistician friends!

Looking at the rows for your two variables of interest will allow you to see how correlated they are with all the numeric variables in your data set. At a glance, it will also show you any large correlation between these other variables. The strength of the correlation with the cause and effect of interest can then help us determine what to do with a given variable.

How strong is strong? It depends. Remember that our goal is to correctly measure the causal effect of our cause of interest on our effect of interest; as a rule of thumb, you can consider as "strong" any correlation that is of the same order of magnitude (i.e., the same number of zeros between the comma and the first nonzero digit) as the correlation between your cause of interest and your effect of interest.

As you can see in Figure 4-13, the correlation coefficient between our two variables of interest is 0.16. The first column indicates the correlations with *NRDeposit* and the second column the correlations with *IsCanceled*. *PreviousCancellation* has correlation coefficients with our variables of interest that are of the same order of magnitude (0.15 and 0.13, respectively). Similarly, *ADR* has a correlation coefficient with *IsCanceled* that is significant by that criterion (0.13).

The "order of magnitude" threshold for inclusion is not scientific in the least, and it can be tightened or loosened depending on how many variables you have at hand. If you have few variables passing the threshold and some other variables are close to it, it is perfectly fine to include them.

You may object that a variable could have a low correlation with either of our variables of interest but still be a confounder that needs to be accounted for. This is true, and you can include a variable even if it is only feebly correlated with our variables of interest, based on a strong theoretical rationale. However, for practical purposes, you should usually focus on variables having at least a moderate level of correlation with our variables of interest.

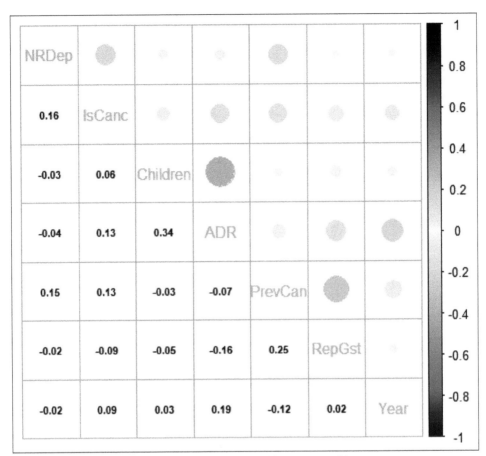

*Figure 4-13. Correlation matrix for numeric and binary variables*

If we include all the correlations that are 0.1 or above in absolute value in Figure 4-13 and exclude the others, our CD is now as in Figure 4-14.

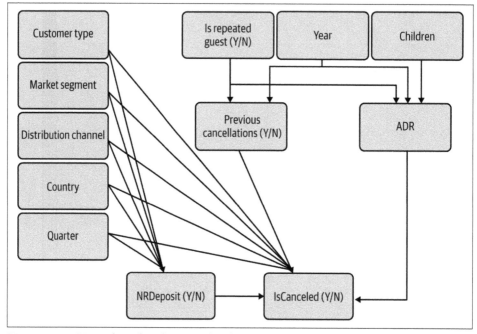

*Figure 4-14. CD with updated arrows for the numeric and binary observable variables*

While the correlation matrix only gives us symmetrical coefficients that could represent arrows in either direction, I have used some common sense and business knowledge to assume the direction of the arrows. The hotel company has no mastery of time, so we can assume that *Years* is the cause and not the effect of variables it's correlated with, although that effect may go through intermediary variables such as trends in society over time. *IsRepeatedGuest* is a prerequisite for *PreviousCancellation*; because it refers to a past event, it must also be the cause of *ADR* or they both share a common cause.

Don't forget that this is a *tentative* CD:

- Some of these correlations are likely false positives (the coefficient appears stronger than it really is, out of sheer randomness) and conversely, some of the smaller correlations may be false negatives.

- At this stage, we're tentatively treating correlations as evidence of causation. Some of the arrows we've drawn in Figure 4-14 may themselves reflect relationships that are confounded. After adequately measuring the relationship between *NRDeposit* and *IsCanceled*, we may want or need to do the same for other relationships (e.g., the one between *IsRepeatedGuest* and *ADR*).

# Relationships Between Categorical Variables

The same logic applies for categorical variables, with the only complication that we can't use Pearson's correlation coefficient. However, a variant of it, Cramer's V (*https://oreil.ly/KAoIa*), has been developed for categorical variables. In R, it is implemented in the rcompanion package:

```
## R
> with(dat, rcompanion::cramerV(NRDeposit, IsCanceled))
Cramer V
   0.165
```

You can see that in the case of binary variables, it yields a result that is quite close to the direct application of Pearson's correlation coefficient. Unfortunately, it is not implemented in Python, but I provide a function to calculate it:

```
## Python
def CramerV(var1, var2):
    ...
    return V

V = CramerV(dat_df['NRDeposit'], dat_df['IsCanceled'])
print(V)
0.16483946381640308
```

Figure 4-15 shows the corresponding correlation matrix, after renaming the variables for readability:

```
## R
dat <- dat %>%
  rename(CustTyp= CustomerType) %>%
  rename(DistCh = DistributionChannel) %>%
  rename(RepGst = IsRepeatedGuest) %>%
  rename(MktSgmt = MarketSegment) %>%
  rename(IsCanc = IsCanceled) %>%
  rename(PrevCan = PreviousCancellations) %>%
  rename(NRDep = NRDeposit)

## Python
dat_df.rename(columns=
              {"CustomerType": "CustTyp",
               "DistributionChannel": "DistCh",
               "IsRepeatedGuest": "RepGst",
               "MarketSegment": "MktSgmt",
               "IsCanceled": "IsCanc",
               "PreviousCancellations": "PrevCan",
               "NRDeposit": "NRDep"},
              inplace=True)
```

This correlation yields a variety of insights. Looking at the bottom row, we can see that *Quarter* is not meaningfully correlated with anything else. This suggests that seasonality is not a relevant factor for our analysis. Conversely, it is possible that a quarter is too coarse a unit of time and that we need to zoom in on very specific time periods such as the Christmas holidays. We can drop *Quarter* from our CD and replace it with an unobserved variable *Seasonality* as a cue for future research.

Our three variables for customer segments, *CustomerType*, *MarketSegment*, and *DistributionChannel*, show a mixed pattern, with some very strong and some weak correlations between them. Similarly, their correlations with other variables are all over the place: all three of them have correlations with *Country* in the 0.1X digits for example, but two of them have high correlations with *RepeatedGuest* (0.35 and 0.4) whereas the third one has a correlation of only 0.11. This suggests that these variables are not just interchangeable, but that they are capturing some aspects of the same behaviors. This calls for further investigation and most likely creating new variables.

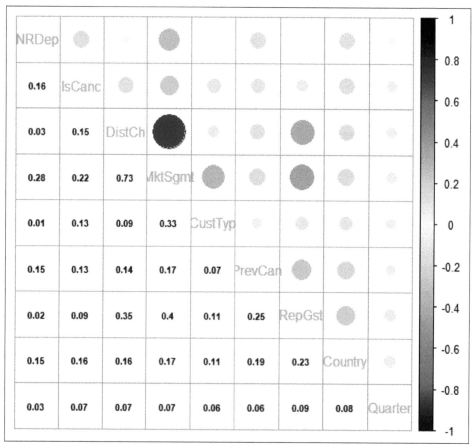

*Figure 4-15. Correlation matrix for categorical and binary variables*

Applying these insights and the same criterion of including only correlations above 0.1, our CD now looks like Figure 4-16.

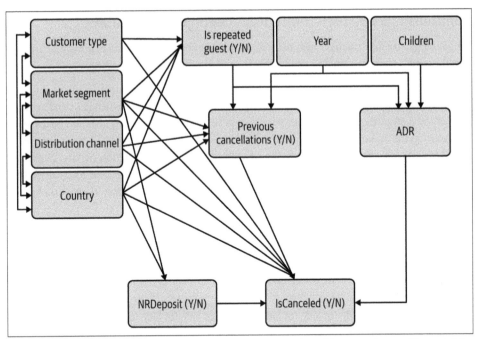

*Figure 4-16. CD with updated arrows for the categorical and binary observable variables*

Our CD is starting to get moderately complex, but it can for the most part be summarized in a few behavioral arguments:

- Our four variables on the left reflect personal characteristics and they are significantly correlated with each other. I have chosen to reflect these correlations with double-headed arrows because trying to determine the direction of arrows would be pointless: *CustomerType* doesn't cause *MarketSegment* any more than the other way around. In reality, after doing the necessary interviews, we should create new variables that capture the deeper personal characteristics at play.

- Personal characteristics appear to affect our variables of interest, potentially causing some confounding.

- Personal characteristics appear to have affected past behaviors *IsRepeatedGuest* and *PreviousCancellation*. (Again, I'm making assumptions on the direction of the effects based on business knowledge. On the face of it, it seems unlikely that having canceled a previous reservation would make someone change country or market segment.) Once we have clarified the nature of the deeper personal characteristics at play, we may decide to fold these past behaviors under some personal characteristics variables, implicitly creating behavioral personas (e.g., "recurring business traveler (Y/N)").

# Relationships Between Numeric and Categorical Variables

Measuring correlations between numeric and categorical variables is a more cumbersome process than measuring correlations within a homogenous category.

Saying that there is a correlation between a numeric and a categorical variable is equivalent to saying that the values of the numeric variable are different on average across the categories of the categorical variable. We can check if this is the case by comparing the mean of the numeric variable across the categories of the categorical variable. For example, we expect that the financial characteristics of the customer may impact the average daily rate for the reservation. It would be best to explore that relationship after having built better variables for customer segmentation, but for the sake of the argument we can use *CustomerType*:

```
## R (output not shown)
> dat %>% group_by(CustTyp) %>% summarize(ADR = mean(ADR))

## Python
dat_df.groupby('CustTyp').agg(ADR = ('ADR', np.mean))
Out[10]:
                        ADR
CustTyp
Contract            92.753036
Group               84.361949
Transient          110.062373
Transient-Party     87.675056
```

We can see that the average daily rate varies substantially between customer types.

 If you're unsure whether the variations are truly substantial or if they only reflect random sampling errors, you can build confidence intervals for them using the Bootstrap, as explained later in Chapter 7.

In our example, there are two numeric variables whose correlation with categorical variables we may want to check: *ADR* and *Year*. We find that *ADR* varies substantially across customer types but that these are reasonably stable over time, which brings us to our final CD for observable variables (Figure 4-17).

At this point, I'd like to reiterate and expand on my earlier warning: in the process of validating observable variables, I have implicitly assumed that correlation was causation. But maybe these relationships are themselves confounded: the correlation between personal characteristics variables and *PreviousCancellation* might be entirely caused by the relationship between personal characteristics variables and *IsRepeatedGuest*.

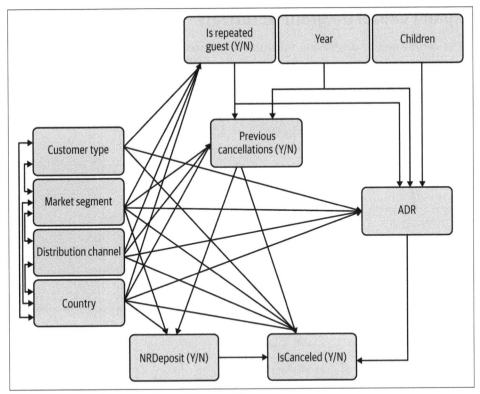

*Figure 4-17. Final CD for observable variables*

Let's imagine for instance that business customers are more likely to be repeated guests. They may then also appear to have a higher rate of previous cancellation than leisure customers even though among repeated guests, business and leisure customers have the exact same rate of previous cancellations.

You can think of these causality assumptions as white lies: they're not true, but it's OK, because we're not trying to build the true, complete CD, we're trying to deconfound the relationship between NRD and cancellation rate. From that perspective, it is *much* more important to get the direction of arrows right than to have unconfounded relationships between variables outside of our variables of interest. If you're still skeptical, one of the exercises in the next chapter explores this question further.

# Expand Causal Diagram Iteratively

After confirming or disproving relationships between observable variables based on data, we have a tentatively complete CD (Figure 4-18).

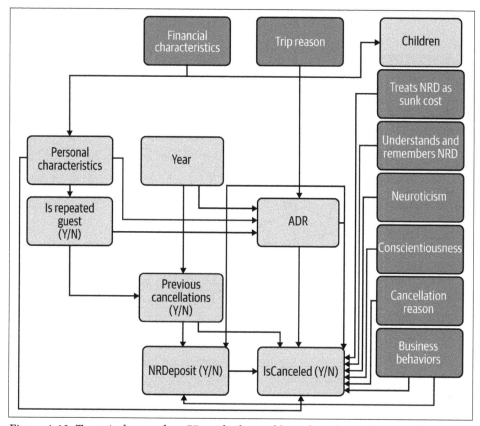

*Figure 4-18. Tentatively complete CD with observable and unobservable variables, grouping personal characteristics variables under one heading for readability*

From there, we'll expand our CD iteratively by identifying proxies for unobserved variables and identifying further causes of the current variables.

## Identify Proxies for Unobserved Variables

Unobserved variables represent a challenge, because even if they are confirmed through interviews or UX research, they can't be accounted for directly in the regression analysis.

We can still try to mitigate them somewhat by identifying potential proxies through interviews and research. For example, we may find that conscientiousness is indeed correlated with a lower rate of cancellation, but also with requesting a confirmation email (Figure 4-19).

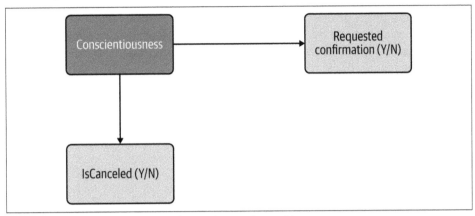

*Figure 4-19. Identifying proxies for unobserved variables*

Of course, requesting a confirmation email is not caused only by conscientiousness—it may also reflect the seriousness of the intent, lack of ease with digital channels, etc. And conversely, it may reduce the cancellation rate by itself, by providing easily accessible information on the reservation. Regardless, if we find that this behavior is negatively correlated with cancellation rate, we may leverage that insight by, for example, sending an SMS reminder to customers who didn't choose to receive a confirmation email.

By brainstorming and validating through research potential proxies for unobserved variables, we're providing meaningful connections between observable variables. Knowing that *RequestedConfirmation* is connected with *IsCanceled* through *Conscientiousness* provides a behavioral rationale for what would otherwise be a raw statistical regularity.

## Identify Further Causes

We'll also expand our CD by identifying causes of the "outer" variables in our CD, i.e., the variables that currently don't have any parent in our CD. In particular, when we have a variable A that affects our cause of interest (possibly indirectly) but not our effect of interest, and another variable B that conversely affects our effect of interest but not our cause of interest, any joint cause of A and B introduces confounding in our CD, because that joint cause is also a joint cause of our two variables of interest.

In our example, the only observable variable without any parent (observable or not) is *Year*, and obviously it can't have one (apart maybe from the laws of physics?), so this step doesn't apply.

## Iterate

As you introduce new variables, you create new opportunities for proxies and further causes. For example, our newly introduced *RequestedConfirmation* may be affected by *Conscientiousness* but also by *TripReason*. This means you should continue to expand your CD until it appears to account for all the relevant variables that you can think of and their interconnections.

There are however significant decreasing returns to this process: as you expand your CD "outward," the newly added variables will tend to have smaller and smaller correlations with your variables of interest, because of all the noise along the way. This means that accounting for them will deconfound your relationship of interest in smaller and smaller quantities.

# Simplify Causal Diagram

The final step once you've decided to stop iteratively expanding the CD is to simplify it. Indeed, you now have a diagram that is hopefully accurate and complete for practical purposes, but it might not be structured in the way that would be most helpful to meet the needs of the business. Therefore I would recommend the following simplification steps:

- Collapse chains when the intermediary variables are not of interest or are unobserved.
- Expand chains when you need to find observed variables or if you want to track the way another variable relates to the diagram.
- Slice variables when you think the individual variables would contain interesting information (e.g., the correlation with one of your variables of interest is really driven by only one slice in particular).
- Combine variables for clarity when reading the diagram or when variation between types does not matter.
- Break cycles wherever you find them by introducing intermediary steps or identifying the aspect of the relationship that is important.

In our example, we may decide that *IsRepeatedGuest*, *Children*, and *Year* do not add value beyond what is captured through *PreviousCancellation* and *ADR*. Indeed, we can dispense with these three variables because they can't confound our relationship of interest (Figure 4-20).

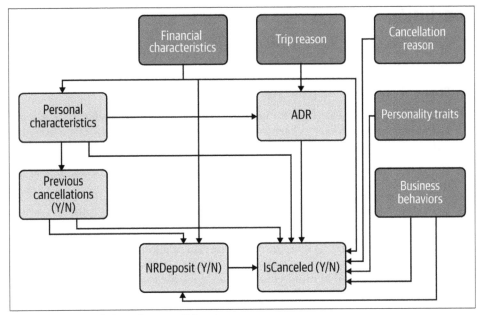

*Figure 4-20. Final CD after simplification*

You should be left with a clean and (somewhat!) readable diagram, although it might be somewhat larger than those we have seen so far.

If this process seems long and somewhat tedious, that's because it is. Understanding your business very well is the price to pay to be able to draw causal inferences on customer (or employee) behavior that are at least somewhat valid when you are not able to run experiments.

Fortunately, this process is extremely cumulative and transferable. Once you've done it for a certain analysis, your knowledge of the causal relationships that matter for your business can be reused for another analysis. Even if you don't go very deep the first time you go through this process, you can just focus on one category of confounders and causes; the next time you run this analysis or a similar one, you can pick up where you left and do a deeper dive on another category, maybe interviewing customers about a different aspect of their experience. Similarly, once someone has gone through the process, a new team member or employee can very easily and quickly acquire the corresponding knowledge and pick up where they left off by looking at the resulting CD or even just the list of relevant variables to keep in mind.

# Conclusion

To use the cliche phrase, building a CD is an art and a science. I have done my best to provide as clear a recipe as possible to do so:

1. Start with the relationship that you're trying to measure.

2. Identify candidate variables to include. Namely, use your behavioral science knowledge and business expertise to identify variables that are likely affecting either of your variables of interest.

3. Confirm which observable variables to include based on their correlations in the data.

4. Iteratively expand your CD by adding proxies for unobserved variables where possible and adding further causes of the variables included so far.

5. Finally, simplify your CD by removing irrelevant relationships and variables.

As you do so, always keep in mind your ultimate goal: measuring the causal impact of your cause of interest on your effect of interest. We'll see in the next chapter how to use a CD to remove confounding in your analysis and obtain an unbiased estimate for that impact. Thus, the best CD is the one that allows you to make the best use of the data you currently have available and that drives fruitful further research.

# CHAPTER 5

# Using Causal Diagrams to Deconfound Data Analyses

Cause-and-effect relationships are so fundamental to our understanding of the world that even a kindergartner intuitively grasps them. However, that intuition—and our data analyses—can be led astray by confounding, as we saw in Chapter 1. If we don't account for a joint cause of our two variables of interest, then we'll misinterpret what's happening and the regression coefficient for our cause of interest will be biased. However, we also saw the risks of taking into account the wrong variables. This makes determining which variables to include or not one of the most crucial questions in deconfounding data analyses and, more broadly, causal thinking.

It is alas a complicated question, with various authors proposing various rules that are more or less expansive. On the more expansive end of the spectrum, you have rules that err towards caution and simplicity—you can think of them as reasoned "everything and the kitchen sink" approaches. On the other end of the spectrum, you have rules that try to zero in on the exact variables required and nothing else, but at the cost of higher complexity and conceptual requirements.

Interestingly, answering that question doesn't require any data. That is, you may want or need data to build the right CD, but once you have a CD that is correct, you don't need to look at any data to identify confounding. This puts us squarely on the CD-to-behaviors edge of our framework (Figure 5-1) and as a consequence we won't use any data in this chapter.

Instead, I will show you two rules for deconfounding with different pros and cons, the "disjunctive cause criterion" and the "backdoor criterion," so that you can choose which one to use depending on your situation. I will set up our business problem in the next section before seeing in turn how to apply the two criteria.

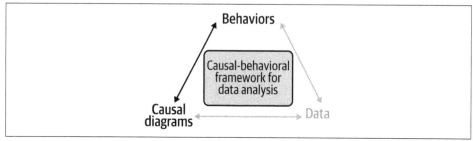

*Figure 5-1. This chapter deals with the relationship between behaviors and causal diagrams*

# Business Problem: Ice Cream and Bottled Water Sales

Our starting point in this chapter is that C-Mart's marketing department has released an internal report titled "the healthy customer," which tracked the long-term trend towards healthier products. Based on that report, C-Mart has launched a marketing campaign titled "Would you like spring water with that?" for its fast-food and ice cream concessions. Our analytical goal is to obtain an unbiased estimate for the impact of ice cream sales on bottled water sales.

By leveraging existing data and dedicated surveys, the marketing analytics team has established the following causal diagram, with our relationship of interest in bold (Figure 5-2).

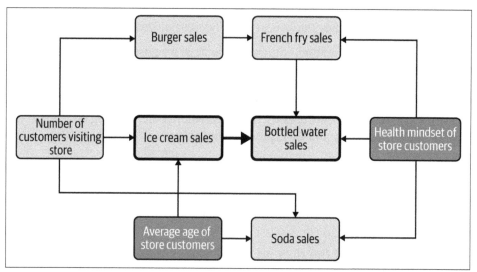

*Figure 5-2. The CD for our business situation*

This CD is moderately complex and it's not immediately obvious where confounders might lurk, so let's break it down into more manageable pieces (Figure 5-3).

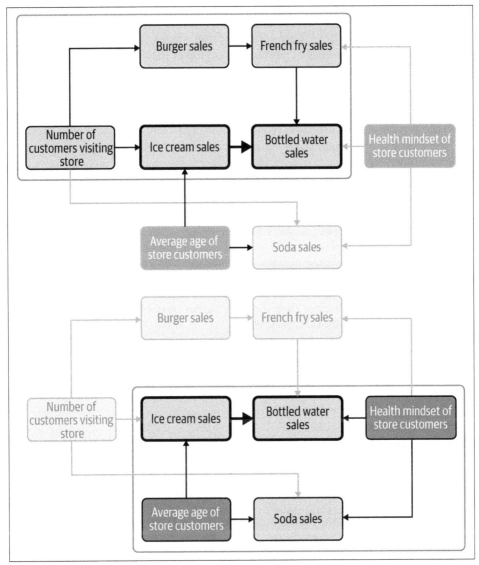

*Figure 5-3. Breaking down our CD into conceptual blocks*

These two conceptual blocks are just a pedagogical tool for understanding: they are not exclusive (our relationship of interest is part of both) nor are they exhaustive (some arrows are in neither of them).

Our first block, in the upper left corner of the CD, shows the connections between the sales of ice cream and the sales of burgers and french fries through the number of customers. In stores with more traffic, and on busier days, overall sales tend to be higher, which makes the various variables move in concert. In addition, store staff has been instructed to offer the "Would you like spring water with that?" prompt both for sales of ice cream and for sales of french fries, on top of the "Would you like fries with that?" prompt for the sales of burgers.

Our second block, in the bottom right corner of the CD, shows the influence of two factors that have been identified in surveys but are not available at the individual sale level: the average age of customers (younger customers, and customers with children, are more likely to purchase sugary products) and the health mindset of customers (health-minded customers are more likely to buy water and less likely to buy sodas, everything else being equal).

 In real-life settings, feel free to break down a large or complex CD into blocks as you see fit for your analyses. This is perfectly fine as long as you do some housekeeping at the end and check the paths from your cause of interest to your effect of interest that are not part of any block: you must make sure that they are not generating confounding, either because they are not confounding in the first place, or because their confounding has been taken care of when analyzing the conceptual blocks.

In this situation, it's not immediately clear whether our relationship of interest between *IceCreamSales* and *BottledWaterSales* is subject to confounding and if so how to resolve it. Technically speaking, we don't have any joint cause of both in our CD. Let's turn to our decision rules.

# The Disjunctive Cause Criterion

The disjunctive cause criterion is our first decision rule for deconfounding. Like an overprotective parent, it goes above and beyond what's strictly necessary to remove confounding, which makes it simpler to understand and apply.

## Definition

The disjunctive cause criterion (DCC) states that:

> Adding to our regression all variables that are a direct cause of both or either of our variables of interest, except mediators between them, removes any confounding of our relationship of interest.

# First Block

Let's start by breaking down this definition based on the first block in our ice cream example:

*1. All variables that are a direct cause of both or either of our variables of interest*
> This means we should include any variable that is only a direct cause of *Ice-CreamSales*, such as *NumberOfCustomers*. We should also include any variable that is only a cause of *BottledWaterSales*, such as *FrenchFrySales*. And finally, we should include any cause of both, but we don't have any in this situation.

*2. Except mediators between them*
> Mediators are variables that "transmit" the impact of our cause of interest on our effect of interest. That is, they are a child of our cause of interest and a parent of our effect of interest. We'll look at mediators in more detail in Chapter 12, so for now I'll just point out that we need to *exclude* them from our list of controls, because including them would cancel some of the causal relationships we're trying to capture. We don't have mediators between *IceCreamSales* and *BottledWaterSales* (i.e., a variable that would be a child of the former and a parent of the latter), so we're good on that front.

*3. Removes any confounding of our relationship of interest*
> If we include the variables described in point 1 but not those described in point 2, then our regression coefficient for the effect of *IceCreamSales* on *BottledWaterSales* will be unconfounded with respect to the variables in our first block.

It's important to note that the DCC is a *sufficient* but not *necessary* rule: applying it is enough to remove confounding but you don't need to. For example, if we have a variable that is only a cause of one of our variables of interest and we are certain it has absolutely no connection whatsoever with any other variable, then it cannot be a confounder and we don't need to include it to remove confounding.

But when you don't have that certainty, the DCC saves you from agonizing over which variable causes which, and what is or isn't a confounder. You may be missing some relationships between variables or believe that there are relationships where there are none; you may believe that a variable is a confounder when it's not or vice versa. As long as you correctly determine whether or not a variable has a direct causal relationship with one of the two variables you're interested in, then you'll make the right decision whether to include it or not.

For example, let's look at the chain from *NumberOfCustomers* to *BottledWaterSales* through *BurgerSales* and *FrenchFrySales*. We saw in Chapter 2 that a chain is a causal diagram that connects variables with arrows in a straight line (Figure 5-4).

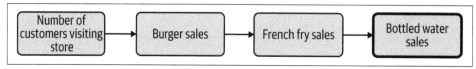

*Figure 5-4. The extended chain in our first block*

Of course, we could represent this chain with the arrows going up, down, or from right to left; the key point is that they are all going in the same direction, which allows us to collapse the chain and treat *NumberOfCustomers* as a direct cause of *BottledWaterSales*. But then *NumberOfCustomers* is indeed a joint direct cause of *IceCreamSales* and *BottledWaterSales*, and a confounder of their relationship (Figure 5-5).

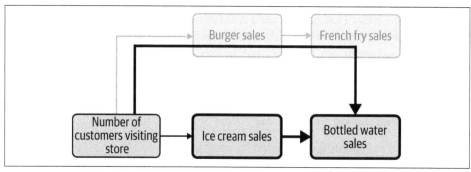

*Figure 5-5. Collapsing the upper chain makes NumberOfCustomers a direct cause of BottledWaterSales*

By definition of the DCC, applying it to this first block means including both *NumberOfCustomers* and *FrenchFrySales* as controls in our regression. Looking at Figure 5-5, we can see that doing so effectively neutralizes the confounding effect of the upper chain. More generally, because chains can be expanded or collapsed at will, a variable that is ultimately a cause of both of our variables of interest (and therefore a confounder) may be hidden behind a succession of intermediary variables in a CD.

The beauty of the DCC is that even if the marketing team had missed the upper chain from *NumberOfCustomers* to *BottledWaterSales* and it was not included in the CD, the requirement of including both *NumberOfCustomers* and *FrenchFrySales* would have taken care of the confounding. On the other hand, based on Figure 5-5, we can see that only including *NumberOfCustomers* would have been enough, and including *FrenchFrySales* too is redundant. This is one of the trade-offs I mentioned in the chapter introduction: the DCC is an expansive rule, which will remove confounding even when there are errors in your CD but at the cost of redundancy and requiring more data. Let's now turn to the second block in our CD.

## Second Block

The second block has more complex relationships between variables (Figure 5-6).

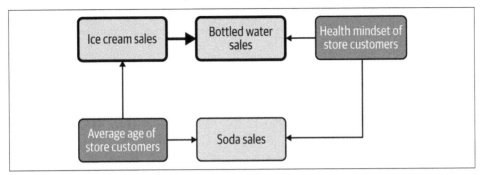

*Figure 5-6. Second block*

Here, the only variable for which we have data beyond our variables of interest is *SodaSales*. It is not a cause of either *IceCreamSales* or *BottledWaterSales*, so the DCC would not call for its inclusion in the regression. However, it would call for the inclusion of both *AverageCustomerAge* and *CustomerHealthMindset*, for which we don't have data. This doesn't necessarily mean that confounding is happening, but we can't be sure that it isn't. This is the biggest limitation of the DCC: if you don't have data on some of the causes of your variables of interest, it cannot help you. Let's now turn to the backdoor criterion.

# The Backdoor Criterion

The backdoor criterion (BC) constitutes an alternative rule to control for confounders. It offers very different trade-offs compared to the disjunctive cause criterion: it is more complex to understand and requires having a fully accurate CD but it zeroes in on actual confounders and does not require including any redundant variable in our regression. In formal terms, it is necessary and sufficient to control for the variables identified by this rule to remove confounding.

# Definitions

The backdoor criterion states that:

> The causal relationship between two variables is confounded if there is at least one unblocked noncausal path between them starting with an arrow to our cause of interest.

> Conversely, to remove all confounding, we need to block all noncausal paths between them starting with an arrow to our cause of interest.

To understand this definition, we'll need to introduce or recall a variety of secondary definitions, in the context of the CD for our example that we'll repeat here (Figure 5-7).

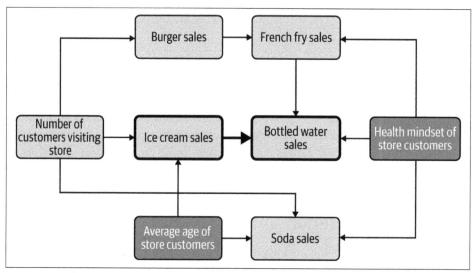

*Figure 5-7. The CD for our business situation*

First, let's recall the definition of a *path*: we say that there is a path between two variables if there are arrows between them, regardless of the direction of the arrows, and if no variable appears twice along the way. A chain is a path along three or more variables, but so are forks and colliders. In that sense, any two variables in a CD are connected by at least one path, and generally several.

For example, there are seven distinct paths between *NumberOfCustomers* and *BottledWaterSales* in our CD:

- *NumberOfCustomers → IceCreamSales → BottledWaterSales*
- *NumberOfCustomers → BurgerSales → FrenchFrySales → BottledWaterSales*
- *NumberOfCustomers → BurgerSales → FrenchFrySales ← CustomerHealthMindset → BottledWaterSales*

- *NumberOfCustomers → BurgerSales → FrenchFrySales ← CustomerHealthMindset → SodaSales ← AverageCustomerAge → IceCreamSales → BottledWaterSales*

- *NumberOfCustomers → SodaSales ← CustomerHealthMindset → BottledWaterSales*

- *NumberOfCustomers → SodaSales ← CustomerHealthMindset → FrenchFrySales → BottledWaterSales*

- *NumberOfCustomers → SodaSales ← AverageCustomerAge → IceCreamSales → BottledWaterSales*

Note that *NumberOfCustomers → BurgerSales → FrenchFrySales ← CustomerHealthMindset → SodaSales ← NumberOfCustomers → IceCreamSales → BottledWaterSales* is not a path because the variable *NumberOfCustomers* appears twice in it, which is forbidden.

A path is *causal* if it's a chain, i.e., all the arrows in it go in the same direction. The label "causal" refers to the fact that a path between two variables is causal if one of the two variables causes the other through that path.

Paths 1 and 2 in the preceding list are causal: they are chains and represent channels through which *NumberOfCustomers* affects *BottledWaterSales*. The other paths are *noncausal* because they each include at least one collider or fork. Remember that a collider is the situation when two variables cause the same one, whereas a fork is the situation when two variables are caused by the same one. For example, paths 3 and 4 both have a collider around *FrenchFrySales*, and path 4 also includes a collider around *SodaSales* and two forks around *CustomerHealthMindset* and *AverageCustomerAge*.

Finally, we'll say that a path between two variables in our CD is *blocked* if either:

- One of the intermediary variables along that path is included in our regression and it's not a collider, or

- In that path there is a collider whose central variable is not included in our regression.

Otherwise, that path is *unblocked*.

The concept of blocked or unblocked is hard to understand because it really encapsulates two different things: whether a path is confounding or not in itself, and whether it's controlled for in our regression. You can think of unblocked as {confounding *and* uncontrolled}, and blocked as {nonconfounding *or* controlled}.

The ultimate root cause of confounding is always a joint cause (Figure 5-8, left panel). However, because we can collapse or expand chains at will, that confounder may be "hidden" behind a number of intermediary variables (Figure 5-8, middle panel). However, we can't collapse a collider in the middle of a chain because it breaks down the direction of the arrows (Figure 5-8, right panel). Therefore a collider blocks confounding, unless we include it in our regression, which neutralizes it.

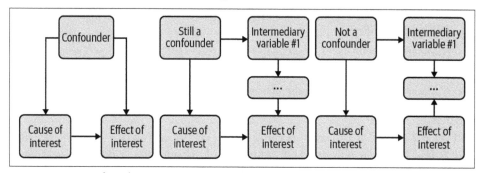

*Figure 5-8. A confounder is a joint cause (left panel), but it can be hidden behind intermediary variables (middle panel), while a collider prevents us from collapsing the chain and therefore removes confounding (right panel)*

## First Block

Now that we've seen the definition of the BC, let's see how it applies to the variables in the first block of the causal diagram. Remember that the DCC required that we include both *NumbersOfCustomers* and *FrenchFrySales* as controls in our regression.

We can start our application of the BC with the condition "starting with an arrow to our cause of interest," which simply means all the causes of our cause of interest, which in this case is *IceCreamSales*. There is only one of them in the first block, namely *NumberOfCustomers*.

For each path through *NumberOfCustomers*, let's apply the other conditions of the BC. The first path from *IceCreamSales* to *BottledWaterSales* through *NumberOfCustomers* within the first block is *IceCreamSales ← NumberOfCustomers → BurgerSales → FrenchFrySales → BottledWaterSales*. This is the path that the DCC caught and controlled by including *NumberOfCustomers* and *FrenchFrySales* in our regression. Let's check the conditions:

- Is that path *noncausal*? Yes, because of the fork around *NumberOfCustomers*.
- Is that path blocked by default? No, as there is no collider in that path, and we haven't included any variable as control yet.

Therefore, this path is confounding our relationship of interest and we need to control for that path by including any one of its noncollider variables in our regression. That is, the BC is telling us that including any of (*NumberOfCustomers*, *BurgerSales*, *FrenchFrySales*) is sufficient to control for that path. However, I have a personal recommendation as to which variable to pick: whenever you can include the first variable along that path, i.e., the cause of your cause of interest, you should do so. In our example, that would be *NumberOfCustomers*. The reason for this choice is that it will also automatically control any other confounding path starting with that variable, which means we don't even have to check any other path starting with that variable.

As you can see, the BC is more economical than the DCC, by leveraging the assumption that we have a complete and correct causal diagram for the variables in that block: whereas the DCC asked us to include *both NumberOfCustomers* and *FrenchFrySales*, the BC only requires including *NumberOfCustomers*, and we can leave the first block without checking any other path.

## Second Block

Remember that the DCC kept mum about the variables in the second block: we couldn't include the variables *AverageCustomerAge* and *CustomerHealthMindset* because we don't have the corresponding data and therefore we were left uncertain whether there was uncontrolled confounding there. The BC will allow us to be much more definitive and precise.

*AverageCustomerAge* is a cause of our cause of interest, *IceCreamSales*, so let's examine the path *IceCreamSales ← AverageCustomerAge → SodaSales ← CustomerHealthMindset → BottledWaterSales*:

- That path is noncausal (i.e., not a chain): it has a fork around *AverageCustomerAge*, a collider around *SodaSales*, and then another fork around *CustomerHealthMindset*.
- Is it blocked by default? Yes, because of the collider around *SodaSales*.

In other words, this path is not confounding and we don't need to control for it in our regression. More than that, including *SodaSales* in our regression would actually create confounding, by unblocking this path!

This configuration of two forks around a collider is peculiar enough that it has a name: the M-pattern, which we can see by rearranging our CD (Figure 5-9). Admittedly, this example might seem unduly contrived. But in case you felt it was artificial and unrealistic, note that it was adapted from an example in *The Book of Why* about actual tobacco litigation in 2006, where the inclusion of a control for seatbelt usage biased the estimated impact of smoking on lung cancer.

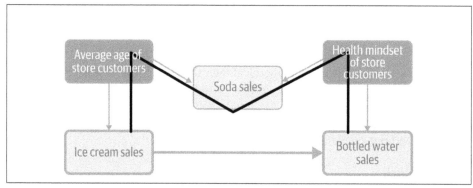

*Figure 5-9. Visualizing the M-pattern in our CD*

Beyond that, because all paths from *IceCreamSales* to *BottledWaterSales* through *AverageCustomerAge* also go through *SodaSales*, they will all be blocked as long as we don't include *SodaSales* in our regression.

Finding confounding in a CD is science: apply the rules and you'll know. But it also has its back-alley shortcuts: having identified two causes of *IceCreamSales* through which confounding was a possibility and having made sure that any confounding would be blocked, we don't have to check each and every path to *IceCreamSales* coming through these two causes. As you build and manipulate more CDs, you'll learn to develop an intuition for it. And if you ever have a doubt, you can always check the rules for each possible path to make sure that you're correct.

The backdoor criterion is more precise than the disjunctive cause criterion, but it's also less robust to errors in our CD. Let's imagine for the sake of the argument that the marketing team made a mistake in building the CD and wrongly concluded that *SodaSales* causes *CustomerHealthMindset* and not the other way around (in that particular case, this doesn't make much sense from a behavioral perspective, but bear with me), leading to the relationships represented in Figure 5-10.

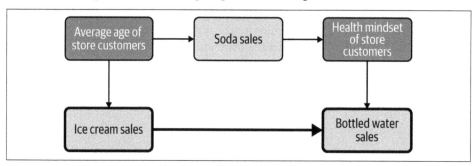

*Figure 5-10. What the second block would look like with an error*

In that case, the BC would lead us to mistakenly believe there is confounding at play, and to include *SodaSales* in our regression, introducing confounding where there had been none.

To recap, the BC identified two potential avenues for confounding, through the two direct causes of *IceCreamSales*. By including *NumberOfCustomers* in our regression, we take care of all potential confounding paths through it. On the other hand, by *not* including *SodaSales*, we leave alone a collider that takes care of any confounding through *AverageCustomerAge*.

## Conclusion

Deconfounding causal relationships is one of the core issues of behavioral data analysis, and the solution to the "correlation is not causation" quagmire. In this chapter, we saw two deconfounding rules, the *disjunctive cause criterion* and the *backdoor criterion*. The first one takes the stance of including all direct causes of our variables of interest (except mediators). The second one is more surgical in its application and gets to the very mechanism of confounding but is less intuitive and less robust to errors in a CD.

# PART III

# Robust Data Analysis

Ideal data is large, complete, and regularly shaped (e.g., normally distributed in the case of numeric variables). This is the data you see in introductory stats courses. Real-life data is often less accommodating, especially when dealing with behavioral data.

In Chapter 6, we'll see how to handle missing data. While missing data is a common occurrence in data analysis, behavioral data adds a layer of complexity: which values are missing is often correlated with individual characteristics and behaviors, and that introduces bias in our analyses. Fortunately, using CDs will allow us to identify and resolve such situations as well as possible.

In Chapter 7, we'll talk about a type of computer simulation called the Bootstrap. It's a very versatile tool that is particularly well suited to behavioral data analysis: it allows us to appropriately measure uncertainty around our estimates when dealing with small or weirdly shaped data. Moreover, when designing and analyzing experiments, it offers an alternative to p-values that will make our lives much simpler.

# CHAPTER 6

# Handling Missing Data

Missing data is a common occurrence in data analysis. In the age of big data, many authors and even more practitioners treat it as a minor annoyance that is given scant thought: just filter out the rows with missing data—if you go from 12 million rows to 11 million, what's the big deal? That still leaves you with plenty of data to run your analyses.

Unfortunately, filtering out the rows with missing data can introduce significant biases in your analysis. Let's say that older customers are more likely to have missing data, for example because they are less likely to set up automated payments; by filtering these customers out you would bias your analysis toward younger customers, who would be overrepresented in your filtered data. Other common methods to handle missing data, such as replacing them by the average value for that variable, also introduce biases of their own.

Statisticians and methodologists have developed methods that have much smaller or even no bias. These methods have not been adopted broadly by practitioners yet, but hopefully this chapter will help you get ahead of the curve!

The theory of missing values is rooted in statistics and can easily get very mathematical. To make our journey in this chapter more concrete, we'll work through a simulated data set for AirCnC. The business context is that the marketing department, in an effort to better understand customer characteristics and motivations, has sent out by email a survey to a sample of 2,000 customers in three states and collected the following information:

- Demographic characteristics
  — Age
  — Gender
  — State (selecting customers from only three states, which for convenience we'll call A, B, and C)
- Personality traits
  — Openness
  — Extraversion
  — Neuroticism
- Booking amount

To keep things simpler, we'll assume that the demographic variables are all causes of booking amount and unrelated to each other (Figure 6-1).

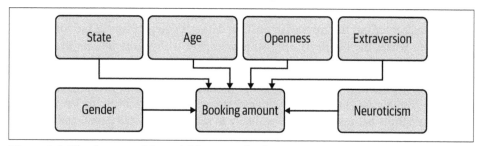

*Figure 6-1. The demographic variables cause booking amount*

 As we discussed in Chapter 2, when I say that demographic variables such as *Gender* and *Extraversion* are causes of *BookingAmount*, I mean two things: first that they are exogenous variables (i.e., variables that are primary causes for our purposes), and second that they are predictors of *BookingAmount* because of causal effects that are heavily mediated as well as moderated by social phenomena.

For instance, the effect of *Gender* is probably mediated by the person's income, occupation, and family status, among many others. In that sense, it would be more accurate to say that *Gender* is a cause of causes of *BookingAmount*. However, it is important to note that this effect is *not* confounded, and as such it is truly causal.

The flow of this chapter will follow the steps you would take when facing a new data set: first, we'll visualize our missing data, to get a rough sense of what's going on. Then, we'll learn how to diagnose missing data and see the classification developed by the statistician Donald Rubin, which is the reference in the matter. The last three sections will show how to handle each one of the categories in that classification.

Unfortunately for Python users, the excellent R packages that we'll be using don't have direct Python counterparts. I'll do my best to show you alternatives and workarounds in Python, but the code will be significantly longer and less elegant. Sorry!

# Data and Packages

One of the luxuries of using simulated data is that we know the true values for the missing data. The GitHub folder for this chapter (*https://oreil.ly/BehavioralDataAnalysisCh6*) contains three data sets (Table 6-1):

- The complete data for our four variables
- The "available" data where some values are missing for some of the variables
- A secondary data set of auxiliary variables that we'll use to complement our analysis

*Table 6-1. Variables in our data*

|  | Variable description | chap6-complete_data.csv | chap6-available_data.csv | chap6-available_data_supp.csv |
|---|---|---|---|---|
| *Age* | Age of customer | Complete | Complete | |
| *Open* | Openness psychological trait, 0-10 | Complete | Complete | |
| *Extra* | Extraversion psychological trait, 0-10 | Complete | Partial | |
| *Neuro* | Neuroticism psychological trait, 0-10 | Complete | Partial | |
| *Gender* | Categorical variable for customer gender, F/M | Complete | Complete | |
| *State* | Categorical variable for customer state of residence, A/B/C | Complete | Partial | |
| *Bkg_amt* | Amount booked by customer | Complete | Partial | |
| *Insurance* | Amount of travel insurance purchased by customer | | | Complete |
| *Active* | Numeric measure of how active the customer bookings are | | | Complete |

In this chapter, we'll use the following packages in addition to the common ones:

```
## R
library(mice) # For multiple imputation
library(reshape) #For function melt()
library(psych) #For function logistic()
```

```
## Python
from statsmodels.imputation import mice # For multiple imputation
import statsmodels.api as sm # For OLS call in Mice
```

# Visualizing Missing Data

By definition, missing data is hard to visualize. Univariate methods (i.e., one variable at a time) will only get us so far, so we'll mostly rely on bivariate methods, plotting two variables against each other to tease out some insights. Used in conjunction with causal diagrams, bivariate graphs will allow us to visualize relationships that would otherwise be very complex to grasp.

Our first step is to get a sense of "how" our data is missing. The mice package in R has a very convenient function md.pattern() to visualize missing data:

```
## R
> md.pattern(available_data)
    age open gender bkg_amt state extra neuro
368  1    1     1      1      1     1     1    0
358  1    1     1      1      1     1     0    1
249  1    1     1      1      1     0     1    1
228  1    1     1      1      1     0     0    2
163  1    1     1      1      0     1     1    1
214  1    1     1      1      0     1     0    2
125  1    1     1      1      0     0     1    2
120  1    1     1      1      0     0     0    3
 33  1    1     1      0      1     1     1    1
 23  1    1     1      0      1     1     0    2
 15  1    1     1      0      1     0     1    2
 15  1    1     1      0      1     0     0    3
 24  1    1     1      0      0     1     1    2
 24  1    1     1      0      0     1     0    3
 23  1    1     1      0      0     0     1    3
 18  1    1     1      0      0     0     0    4
     0    0     0    175    711   793  1000 2679
```

The md.pattern() function returns a table where each row represents a pattern of data availability. The first row has a "1" for each variable, so it represents complete records. The number on the left of the table indicates the number of rows with that pattern, and the number on the right indicates the number of fields that are missing in that pattern. We have 368 complete rows in our data. The second row has a "0" for *Neuroticism* only, so it represents records where only *Neuroticism* is missing; we have 358 such rows. The numbers at the bottom of the table indicate the number of missing values for the corresponding variables, and the variables are ordered by increasing number of missing values. *Neuroticism* is the last variable to the right, which means it has the highest number of missing values, 1,000. This function also conveniently returns a visual representation of the table (Figure 6-2).

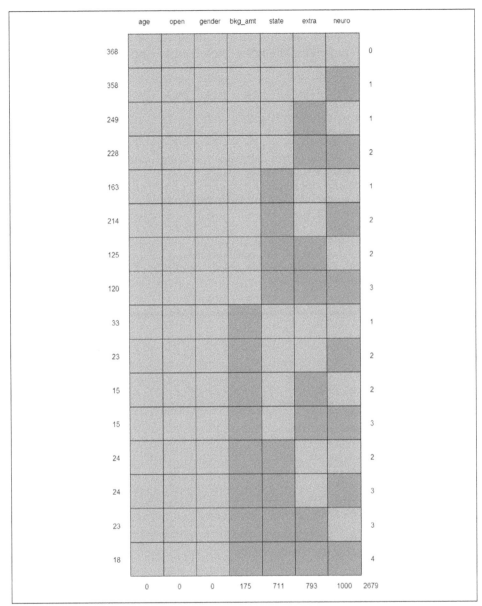

*Figure 6-2. Patterns of missing data*

As we can see in Figure 6-2, the variables *Age*, *Openness*, and *Gender* don't have any missing data, but all the other variables do. We can obtain the same results in Python with an ad hoc function I wrote, although in a less readable format:

```Python
## Python
def md_pattern_fun(dat_df):
    # Getting all column names
    all_cols = dat_df.columns.tolist()
    # Getting the names of columns with some missing values
    miss_cols = [col for col in all_cols if dat_df[col].isnull().sum()]
    if miss_cols == all_cols: dat_df['index'] = dat_df.index
    # Removing columns with no missing values
    dat_df = dat_df.loc[:,miss_cols]
    #Showing total number of missing values per variable
    print(dat_df.isnull().sum())
    # Adding count value
    dat_df['count'] = 1
    # Showing count for missingness combinations
    print(dat_df.isnull().groupby(miss_cols).count())
md_pattern_fun(available_data_df)
```

```
extra        793
neuro       1000
state        711
bkg_amt      175
dtype: int64
```

| extra | neuro | state | bkg_amt | count |
|-------|-------|-------|---------|-------|
| False | False | False | False | 368 |
|       |       |       | True | 33 |
|       |       | True | False | 163 |
|       |       |       | True | 24 |
|       | True | False | False | 358 |
|       |       |       | True | 23 |
|       |       | True | False | 214 |
|       |       |       | True | 24 |
| True | False | False | False | 249 |
|       |       |       | True | 15 |
|       |       | True | False | 125 |
|       |       |       | True | 23 |
|       | True | False | False | 228 |
|       |       |       | True | 15 |
|       |       | True | False | 120 |
|       |       |       | True | 18 |

The output is composed of two tables:

- The first table indicates the total number of missing values for each variable in our data, as seen at the bottom of Figure 6-2. *Extraversion* has 793 missing values, and so on.

- The second table represents the details of each missing data pattern. The variables above the logical values on the left (i.e., *Extraversion, Neuroticism, State, BookingAmount*) are the ones with some missing values in the data. Each line of the table indicates the number of rows with a certain pattern of missing data. The first line is made of four `False`, i.e., the pattern where none of the variables has missing data, and there are 368 such rows in our data, as you've seen previously in the first line of Figure 6-2. The second line only changes the last `False` to `True`, with the first three `False` omitted for readability (i.e., any blank logical value should be read up). This pattern `False/False/False/True` happens when only *BookingAmount* has a missing value, which happens in 33 rows of our data, and so on.

Even with such a small data set, this visualization is very rich and it can be hard to know what to look for. We will explore two aspects:

*Amount of missing data*
How much of our data is missing? For which variables do we have the highest percentage of missing data? Can we just discard rows with missing data?

*Correlation of missingness*
Is data missing at the individual or variable level?

## Amount of Missing Data

The first order of business is to determine how much of our data is missing and which variables have the highest percentage of missing data. We can find the necessary values at the bottom of Figure 6-2, with the number of missing values per variable, in increasing order of missingness, or at the bottom of the Python output. If the amount of missing data is very limited, e.g., you have a 10-million-row data set where no variable has more than 10 missing values, then handling them properly through multiple imputation would be overkill as we'll see later. Just drop all the rows with missing data and be done with it. The rationale here is that even if the missing values are extremely biased, there are too few of them to materially influence the outcomes of your analysis in any way.

In our example, the variable with the highest number of missing values is *Neuroticism*, with 1,000 missing values. Is that a lot? Where is the limit? Is it 10, 100, 1,000 rows or more? It depends on the context. A quick-and-dirty strategy that you can use is as follows:

1. Take the variable with the highest number of missing values and create two new data sets: one where all the missing values are replaced by the minimum of that variable and one where they are replaced by the maximum of that variable.

2. Run a regression for that variable's most important relationship with each of the three data sets you now have. For example, if that variable is a predictor of your effect of interest, then run that regression.

3. If the regression coefficient is not materially different across the three regressions, i.e., you would draw the same business implications or take the same actions based on the different values, then you're below the limit and you can drop the missing data. In simpler words: would these numbers mean the same thing to your business partners? If so, you can drop the missing data.

This rule of thumb is easily applied to numeric variables, but what about binary or categorical variables?

For binary variables, the minimum will be 0 and the maximum will be 1, and that's OK. The two data sets you create translate into a best-case scenario and a worst-case scenario.

For categorical variables, the minimum and maximum rule must be slightly adjusted: replace all the missing values with either the least frequent or the most frequent category.

Let's do that here, for example for *Neuroticism*. *Neuroticism* is a predictor of our effect of interest, *BookingAmount*, so we'll use that relationship as indicated earlier:

```
## R (output not shown)
min_data <- available_data %>%
  mutate(neuro = ifelse(!is.na(neuro), neuro, min(neuro, na.rm = TRUE)))
max_data <- available_data %>%
  mutate(neuro = ifelse(!is.na(neuro), neuro, max(neuro, na.rm = TRUE)))
summary(lm(bkg_amt~neuro, data=available_data))
summary(lm(bkg_amt~neuro, data=min_data))
summary(lm(bkg_amt~neuro, data=max_data))

## Python (output not shown)
min_data_df = available_data_df.copy()
min_data_df.neuro = np.where(min_data_df.neuro.isna(), min_data_df.neuro.min(),
                    min_data_df.neuro)

max_data_df = available_data_df.copy()
max_data_df.neuro = np.where(max_data_df.neuro.isna(), max_data_df.neuro.max(),
                    max_data_df.neuro)


print(ols("bkg_amt~neuro", data=available_data_df).fit().summary())
```

```
print(ols("bkg_amt~neuro", data=min_data_df).fit().summary())
print(ols("bkg_amt~neuro", data=max_data_df).fit().summary())
```

The results are as follows:

- The coefficient based on the available data is −5.9.
- The coefficient based on replacing the missing values with the minimum of *Neuroticism* is −8.0.
- The coefficient based on replacing the missing values with the maximum of *Neuroticism* is 2.7.

These values are very different from each other, to the point of having different signs, therefore we're definitely above the threshold for material significance. We can't simply drop the rows that have missing data for *Neuroticism*. Applying the same approach to the other variables would also show that we can't disregard their missing values and need to handle them adequately.

## Correlation of Missingness

Once we have determined which variables we need to deal with, we'll want to know how much their missingness is correlated. If you have variables whose missingness is highly correlated, this suggests that the missingness of one causes the missingness of others (e.g., if someone stops answering a survey halfway through, then all answers after a certain point will be missing). Alternatively, their missingness may have a common cause (e.g., if some subjects are more reluctant to reveal information about themselves). In both of these cases, identifying correlation in missingness will help you build a more accurate CD, which will save you time and make your analyses more effective.

Let's look at it through a simple illustration: imagine that we have interview data for two offices: Tampa and Tacoma. In both offices, candidates must go through the same mandatory three interview sections, but in Tampa the first interviewer is responsible for recording all the scores of a candidate whereas in Tacoma each interviewer records the score for their section. Interviewers are human beings and they sometimes forget to turn in the data to HR. In Tampa, when an interviewer forgets to turn in the data, we have no data whatsoever for the candidate, apart from their ID in the system (Figure 6-3 shows the data for Tampa only).

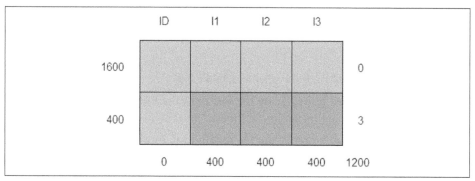

*Figure 6-3. Highly correlated missingness in Tampa data*

The telltale sign for high missingness correlation is a line with a large number of light-shaded squares (here 3) that represent a high number of cases (here 400 out of 2,000 total rows). In addition, the figure has no line with only one or two light squares.

In such a situation, it wouldn't make sense to analyze our missing data one variable at a time. If we find that data for the first section is highly likely to be missing when Murphy is the first interviewer, then it will also be true for the other sections. (You had one job, Murphy!)

In Tacoma, on the other hand, the missingness of the different sections is entirely uncorrelated (Figure 6-4).

The pattern is the opposite of Tampa's:

- We have a high number of lines with few missing variables (see all the 1s and 2s on the right of the figure).

- These lines represent the bulk of our data (we can see on the left that only 17 rows have 3 missing variables).

- Lines with a high number of light squares at the bottom of the figure represent very few cases (the same 17 individuals) because they are the result of independent randomness.

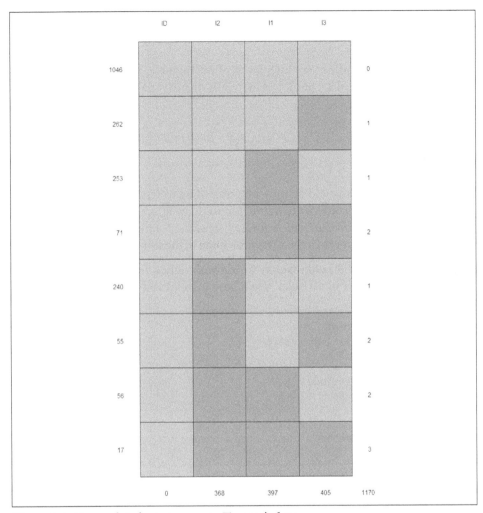

*Figure 6-4. Uncorrelated missingness in Tacoma's data*

The argument for the last bullet point can be extended by looking more broadly at what we could call "Russian dolls" sequences of increasing missingness where each pattern adds a missing variable on the previous pattern, for example (I3) → (I3, I2) → (I3, I2, I1). The corresponding numbers of cases are 262 → 55 → 17. These numbers form a decreasing sequence, which is logical because if the missingness of the variables is completely uncorrelated, we have:

*Prob(I3 missing & I2 missing) = Prob(I3 missing) * Prob(I2 missing)*

*Prob(I3 missing & I2 missing & I1 missing) = Prob(I3 missing) \* Prob(I2 missing) \* Prob(I1 missing)*

With a small sample and/or very high levels of missingness, these equations may not hold exactly true in our data, but if less than 50% of any variable is missing we should generally have:

*Prob(I3 missing & I2 missing & I1 missing) < Prob(I3 missing & I2 missing) < Prob(I3 missing)*

In real-life situations, it would be rather cumbersome to test all these inequalities by yourself, although you could write a function to do so at scale. Instead, I would recommend looking at the visualization for any significant outlier (i.e., a value for several variables much larger than the values for some of the same variables).

More broadly, this visualization is easy to use with only a few variables. As soon as you have a large number of variables, you'll have to build and visualize the correlation matrix for missingness:

```
## R (output not shown)
# Building the correlation matrices
tampa_miss <- tampa %>%
  select(-ID) %>%
  mutate(across(everything(),is.na))
tampa_cor <- cor(tampa_miss) %>%
  melt()

tacoma_miss <- tacoma %>%
  select(-ID) %>%
  mutate(across(everything(),is.na))
tacoma_cor <- cor(tacoma_miss) %>%
  melt()


## Python (output not shown)
# Building the correlation matrices
tampa_miss_df = tampa_df.copy().drop(['ID'], axis=1).isna()
tacoma_miss_df = tacoma_df.copy().drop(['ID'], axis=1).isna()

tampa_cor = tampa_miss_df.corr()
tacoma_cor = tacoma_miss_df.corr()
```

Figure 6-5 shows the resulting correlation matrices. In the one on the left, for Tampa, all the values are equal to 1: if one variable is missing, then the other two are as well. In the correlation matrix on the right, for Tacoma, the values are equal to 1 on the main diagonal but 0 everywhere else: knowing that one variable is missing tells you nothing about the missingness of the others.

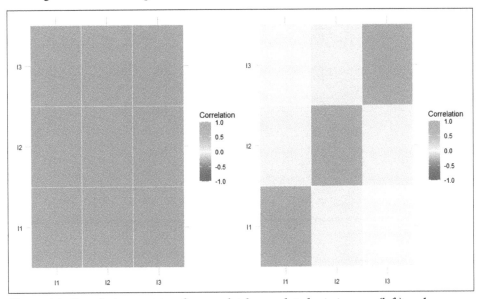

*Figure 6-5. Correlation matrices for completely correlated missingness (left) and completely uncorrelated missingness (right)*

Let's get back to our AirCnC data set and see where it falls between the two extremes outlined in our theoretical interview example. Figure 6-6 repeats Figure 6-2 for ease of access.

Figure 6-6 falls somewhere in the middle: all possible patterns of missingness are fairly represented, which suggests that we don't have strongly clustered sources of missingness.

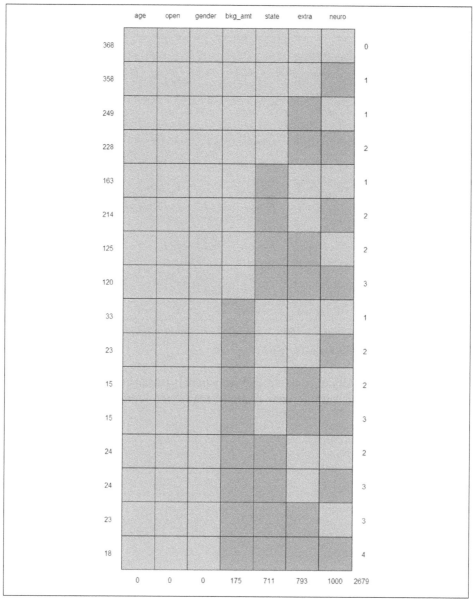

*Figure 6-6. Patterns of missing data (repeats Figure 6-2)*

Figure 6-7 shows the correlation matrix of missingness for our AirCnC data. As you can see, the missingness of our variables is almost entirely uncorrelated, well within the range of random fluctuations. If you want to familiarize yourself more with correlation patterns in missingness, one of the exercises for this chapter on GitHub (*https://oreil.ly/BehavioralDataAnalysisCh6*) asks you to identify a few of them. As a reminder, looking at correlation patterns is never necessary in itself, but it can often be illuminating and save you time.

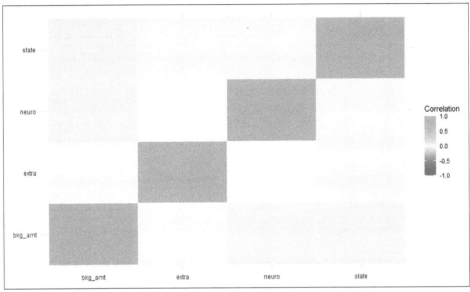

*Figure 6-7. Correlation matrix of missingness in our AirCnC data*

# Diagnosing Missing Data

Now that we have visualized our missing data, it's time to understand what causes it. This is where causal diagrams come in, as we'll use them to represent the causal mechanisms of missing data.

Let's start with a very simple example from Chapter 1. When introducing causal diagrams, I mentioned that unobserved variables, such as a customer's taste for vanilla ice cream, are represented in a darker shaded rectangle (Figure 6-8).

*Figure 6-8. Unobserved variables are represented in a darker-shaded rectangle*

Unobserved variables, which are sometimes called "latent" variables in certain disciplines, refer to information that we don't have in practice, even though it may or may not in theory be accessible. In the present case, let's say that we force our customers to disclose their taste for vanilla before making a purchase. This would create the corresponding data in our systems, which we would then use for our data analyses (Figure 6-9).

*Figure 6-9. Collecting previously unobserved information*

However, it is generally poor business practice to try and force customers to disclose information they don't want to, and it's often left optional. More generally, a large amount of data is collected on some customers but not others. We'll represent that situation by drawing the corresponding box in the CD with dashes (Figure 6-10).

*Figure 6-10. Representing partially observed variables with a dashed box*

For example, with three customers, we could have the following data, with one customer refusing to disclose their taste for vanilla ice cream (Table 6-2).

*Table 6-2. The data underlying our CD*

| Customer name | Taste for vanilla | Stated taste | Purchased IC at stand (Y/N) |
|---|---|---|---|
| Ann | Low | Low | N |
| Bob | High | High | Y |
| Carolyn | High | N/A | Y |

## From Missing Values to Wrong or False Values

In this book, we'll make the simplifying assumption that values are either correct or missing. People never lie, misremember, or make entry errors. Of course, they do so in real life, and that's something you'll have to address by using the insights from this chapter and Chapter 2: assume that there's a hidden variable with the true values, and an observable variable that reflects the hidden variable with some "noise." Then determine if that noise is purely random, dependent on the value of another variable, or dependent on the hidden variable, similar to Rubin's classification (discussed later in this chapter). For example, people who have bought a product they may be embarrassed to confess using (a toupee?) are somewhat likely to pretend they don't, but the converse is highly unlikely. That would make *ToupeePurchases* "false not at random" in Rubin's terminology.

In this chapter, we're interested in understanding what causes missingness of a variable, and not just what causes the values of a variable. Therefore, we'll create a variable to track when the stated taste variable is missing (Table 6-3).

*Table 6-3. Adding a missingness variable*

| Customer name | Taste for vanilla | Stated taste for vanilla | Stated taste missing (Y/N) | Purchased IC at stand |
|---|---|---|---|---|
| Ann | Low | Low | N | N |
| Bob | High | High | N | Y |
| Carolyn | High | N/A | Y | Y |

Let's add that variable to our CD (Figure 6-11).

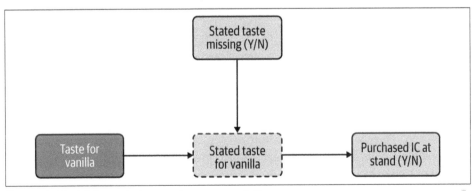

*Figure 6-11. Adding missingness to our causal diagram*

We conventionally make missingness a cause of the corresponding partially observed variable. The intuition is that the information exists fully in the unobserved variable, and the partially observed variable is equal to the unobserved variable, unless information is "hidden" by the missingness variable. This convention will make our lives much easier, because it will allow us to express and discuss causes of missingness in the CD that represent our relationships of interest, instead of having to consider missingness separately.

Now that missingness is part of our CD, the natural next step is to ask ourselves, "What causes it?"

## Causes of Missingness: Rubin's Classification

There are three basic and mutually exclusive possibilities for what causes missingness of a variable, which have been categorized by statistician Donald Rubin.

First, if the missingness of a variable depends only on variables outside of our data, such as purely random factors, that variable is said to be *missing completely at random* (MCAR) (Figure 6-12).

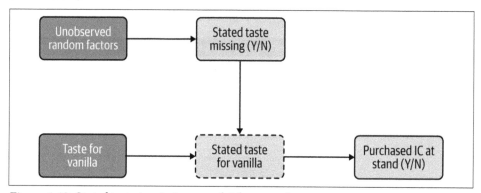

*Figure 6-12. Stated taste is missing completely at random*

Then, a variable goes from MCAR to *missing at random* (MAR) if even one variable in our data affects its missingness. Variables outside of our data and random factors may also play a role, but the value of the variable in question may not affect its own missingness. This would be the case for instance if *Purchased* caused the missingness of *StatedVanillaTaste*, e.g., because we only interview customers who purchased instead of passersby (Figure 6-13).

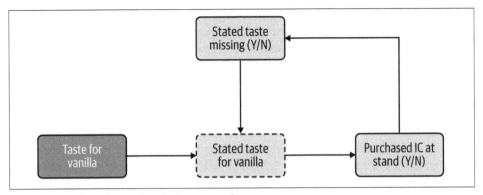

*Figure 6-13. Stated taste is missing at random*

Finally, any variable whose value influences its own missingness is *missing not at random* (MNAR), even if other variables inside or outside of the data also affect the missingness. Other variables inside or outside of our data may also play a role, but a variable goes from MCAR or MAR to MNAR as soon as the variable influences its own missingness. In our example, this would mean that *VanillaTaste* causes the missingness of *StatedVanillaTaste* (Figure 6-14).

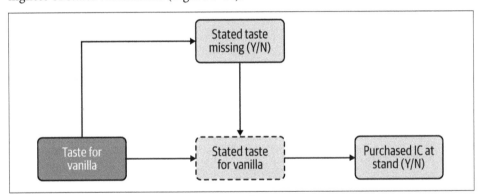

*Figure 6-14. Stated taste is missing not at random*

We represent the idea that the values of a variable influence its missingness by drawing an arrow from the unobserved variable, and not the partially observed variable. This way, we can make meaningful statements such as "all values that are in reality below a certain threshold are missing in our data." If the arrow came from the partially observable variable, we would be stuck with uninformative statements such as "values that are missing cause themselves to be missing."

In an ideal world, the rest of the section would consist of recipes to identify each category of missingness. Unfortunately, missing data analysis is still an open area of research that has not yet been fully explored. In particular, how missingness and causality interact is not well understood. Therefore, dealing with missing data remains more art than science. Trying to create systematic recipes would require dealing with an intractable number of exceptions, as well as introducing circular arguments such as "pattern X indicates that variable 1 is MAR unless variable 2 is MNAR; pattern Y indicates that variable 2 is MNAR unless variable 1 is MAR." I have done my best to cover as many cases as possible within a limited data set, but in the real world you might encounter situations that are "a little bit of this and a little bit of that" and you'll have to make judgment calls as to how to proceed.

Some good news, however, is that with a few exceptions that I'll call out, being cautious takes more time but doesn't introduce bias. When you're uncertain whether a variable is MCAR, MAR, or MNAR, just assume the worst of the possible scenarios and your analyses will be as unbiased as they can be.

With that caveat in mind, let's get back to our AirCnC data and see how we can diagnose missingness in a realistic data set. As a quick refresher, our data set contains the following variables:

- Demographic characteristics
  — Age
  — Gender
  — State (A, B, and C)
- Personality traits
  — Openness
  — Extraversion
  — Neuroticism
- Booking amount

## Diagnosing MCAR Variables

MCAR variables are the simplest case. A sensor went faulty, a bug prevented data transmission from a customer's mobile app, or a customer just missed the field to enter their taste for vanilla ice cream. Regardless, missingness happens in a way that is intuitively "random." We diagnose MCAR variables by default: if a variable doesn't appear to be MAR, we'll treat it as MCAR. In other words, you can think of MCAR as our null hypothesis in the absence of evidence to the contrary.

The main tool that we'll use to diagnose missingness is a logistic regression of whether a variable is missing on all the other variables in our data set. Let's look, for example, at the *Extraversion* variable:

```
## Python (output not shown)
available_data_df['md_extra'] = available_data_df['extra'].isnull().astype(float)
md_extra_mod =smf.logit('md_extra~age+open+neuro+gender+state+bkg_amt',
                  data=available_data_df)
md_extra_mod.fit().summary()

## R
> md_extra_mod <- glm(is.na(extra)~.,
                  family = binomial(link = "logit"),
                  data=available_data)
> summary(md_extra_mod)

...
Coefficients:
             Estimate Std. Error z value Pr(>|z|)
(Intercept) -0.7234738  0.7048598  -1.026    0.305
age         -0.0016082  0.0090084  -0.179    0.858
open         0.0557508  0.0425013   1.312    0.190
neuro        0.0501370  0.0705626   0.711    0.477
genderF     -0.0236904  0.1659661  -0.143    0.886
stateB      -0.0780339  0.2000428  -0.390    0.696
stateC      -0.0556228  0.2048822  -0.271    0.786
bkg_amt     -0.0007701  0.0011301  -0.681    0.496
...
```

None of the variables has a large and strongly statistically significant coefficient. In the absence of any other evidence, this suggests that the source of missingness for *Extraversion* is purely random and we'll treat our *Extraversion* variable as MCAR.

You can think of MCAR data as rolling dice or flipping a coin. Both of these actions are "random" from our perspective, but they still obey the laws of physics. Theoretically, if we had enough information and computing power, the outcome would be entirely predictable. The same could happen here. By saying that *Extraversion* is MCAR, we're not saying "the missingness of *Extraversion* is fundamentally random and unpredictable," we're just saying "none of the variables *currently included* in our analysis is correlated with the missingness of *Extraversion*." But maybe—and even probably—other variables (conscientiousness? trust? familiarity with technology?) would be correlated. Our goal is not to make a philosophical statement about *Extraversion*, but to determine if its missingness may bias our analyses, given the data currently available.

## Diagnosing MAR Variables

MAR variables are variables whose missingness depends on the values of other variables in our data. If other variables in our data set are predictive of a variable's missingness, then MAR becomes our default hypothesis for that variable, unless we have strong enough evidence that it's MNAR. Let's see what this looks like with the *State* variable:

```
## R (output not shown)
md_state_mod <- glm(is.na(state)~.,
                    family = binomial(link = "logit"),
                    data=available_data)
summary(md_state_mod)

## Python
available_data_df['md_state'] = available_data_df['state'].isnull()\
    .astype(float)
md_state_mod =smf.logit('md_state~age+open+extra+neuro+gender+bkg_amt',
                    data=available_data_df)
md_state_mod.fit(disp=0).summary()
...
                coef    std err     z       P>|z|   [0.025  0.975]
Intercept     -0.2410   0.809    -0.298     0.766   -1.826   1.344
gender[T.F]   -0.1742   0.192    -0.907     0.364   -0.551   0.202
age            0.0206   0.010     2.035     0.042    0.001   0.040
open           0.0362   0.050     0.727     0.467   -0.061   0.134
extra          0.0078   0.048     0.162     0.871   -0.087   0.102
neuro         -0.1462   0.087    -1.687     0.092   -0.316   0.024
bkg_amt       -0.0019   0.001    -1.445     0.149   -0.005   0.001
...
```

*Age* is mildly significant, with a positive coefficient. In other words, older customers appear less likely to provide their state. The corresponding causal diagram is represented in Figure 6-15.

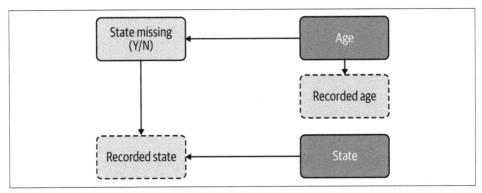

*Figure 6-15. Gender missing at random*

We can confirm that correlation by plotting the density of *State* missingness by recorded *Age* (Figure 6-16). *State* has more observed values for younger customers than for older customers, or conversely, more missing values for older customers than for younger customers.

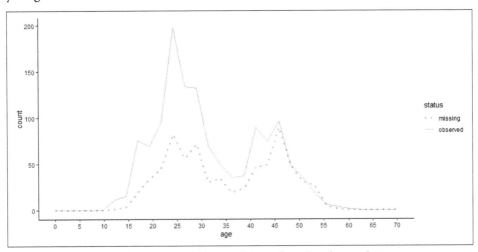

*Figure 6-16. Density of missing and observed State data, by observed Age*

One limitation of this density plot is that it doesn't show the rows where the X variable (here *Age*) is also missing. This could be problematic or misleading when that variable also has missing values. A possible trick is to replace the missing values for the X variable by a nonsensical value, such as −10. *Age* doesn't have any missing value, so instead we'll use as our X variable *Extraversion*, which has missing values. Let's plot the density of observed and missing *State* data, by values of *Extraversion* (Figure 6-17).

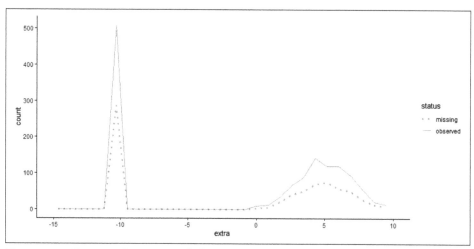

*Figure 6-17. Density of missing and observed State data by level of Extraversion, including missing Extraversion*

Figure 6-17 shows that among individuals with no data for *Extraversion*, there are disproportionately more individuals for which we observe *State* than individuals for which *State* is missing. Overall, we're seeing strong evidence that *State* is not MCAR but indeed MAR, because its missingness appears correlated with other variables available in our data set.

 You may have noticed that I used the word "correlation" earlier when talking about the relationship between *Age* (or *Extraversion*) and the missingness of *State*. Indeed, we've only shown correlation so far, and it is entirely possible that *Age* doesn't cause missingness of *State* but that they are both caused by a third unobserved variable. Fortunately, when talking about missingness, the causal nature of a correlation (or lack thereof) doesn't affect our analyses. Loosely equating the two will not introduce bias because we'll never actually deal with the coefficient for that relationship.

## Diagnosing MNAR Variables

MNAR variables are variables whose missingness depends on their own values: higher values are more likely to be missing than lower values, or vice versa. This situation is both the most problematic for data analysis and the trickiest to diagnose. It is trickiest to diagnose because, by definition, we don't know the values that are missing. Therefore, we'll need to do a bit more sleuthing.

Let's look at the *Neuroticism* variable, and as before, start by running a regression of its missingness on the other variables in the data:

```python
## Python (output not shown)
available_data_df['md_neuro'] = available_data_df['neuro'].isnull()\
    .astype(float)
md_neuro_mod =smf.logit('md_neuro~age+open+extra+state+gender+bkg_amt',
                        data=available_data_df)
md_neuro_mod.fit(disp=0).summary()
```

```r
## R
md_neuro_mod <- glm(is.na(neuro)~.,
                    family = binomial(link = "logit"),
                    data=available_data)
summary(md_neuro_mod)
```

```
...
Coefficients:
            Estimate Std. Error z value Pr(>|z|)
(Intercept) -0.162896   0.457919  -0.356  0.72204
age         -0.012610   0.008126  -1.552  0.12071
open         0.052419   0.038502   1.361  0.17337
extra       -0.084991   0.040617  -2.092  0.03639 *
genderF     -0.093537   0.151376  -0.618  0.53663
stateB       0.047106   0.181932   0.259  0.79570
stateC      -0.128346   0.187978  -0.683  0.49475
bkg_amt      0.003216   0.001065   3.020  0.00253 **
...
```

We can see that *BookingAmount* has a strongly significant coefficient. On the face of it, this would suggest that *Neuroticism* is MAR on *BookingAmount*. However, and this is a key clue, *BookingAmount* is a child of *Neuroticism* in our CD. From a behavioral perspective, it also seems more likely that *Neuroticism* is MNAR rather than MAR on *BookingAmount* (i.e., the missingness is driven by a personality trait rather than by the amount a customer spent).

A way to confirm our suspicion is to identify another child of the variable with missing data, ideally as correlated with it as possible and as little correlated as possible with the first child. In our secondary data set, we have data about the total amount of travel insurance that customers purchased over their lifetime with the company. The fee per trip depends on trip characteristics that are only very loosely correlated with the booking amount, so we're good on that front. Adding *Insurance* to our data set, we find that it's strongly predictive of the missingness of *Neuroticism*, and that the distributions of *Insurance* amount with observed and missing *Neuroticism* are vastly different from each other (Figure 6-18).

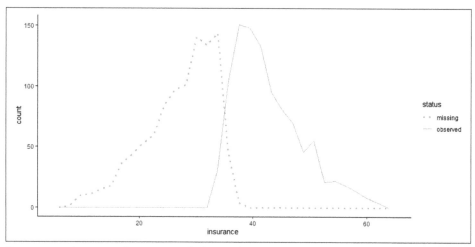

*Figure 6-18. Density of missing and observed Neuroticism data, by observed Insurance amount*

The more children variables we find correlated with the missingness of *Neuroticism*, the stronger our case becomes that the variable is MNAR. As we'll see later, the way to handle a MNAR variable is to add auxiliary variables to our imputation model, and our children variables are perfect candidates for that, so finding several of them is not a waste of time but a head start for the next step.

Technically, we can never fully prove that a variable is MNAR and not just MAR on several of its children, but that is not an issue: auxiliary variables do not bias the imputation if the original variable is indeed MAR and not MNAR.

## Missingness as a Spectrum

Rubin's classification relies on binary tests. For example, as soon as a variable is more likely to be missing for higher values than for lower values (or vice versa), it is MNAR, regardless of any other consideration. However, the shape of that relationship between values and missingness matters for practical purposes: if *all* values of a variable are missing above or below a certain threshold, we'll need to handle this variable differently from our default approach. This situation can also occur with MAR variables so it's worth taking a step back and thinking more broadly about shapes of missingness.

We can think of the missingness of a variable as falling on a spectrum from fully probabilistic to fully deterministic. At the "probabilistic" end of the spectrum, the variable is MCAR and all values are as likely to be missing. At the "deterministic" end of the spectrum, there is a threshold: the values are missing for all individuals on one side of the threshold and available for all individuals on the other side of the threshold. This often results from the application of a business rule. For example, in a hiring context, if only candidates with a GPA above 3.0 get interviewed, you wouldn't have any interview score for candidates below that threshold. This would make *InterviewScore* MAR on *GPA* (Figure 6-19).

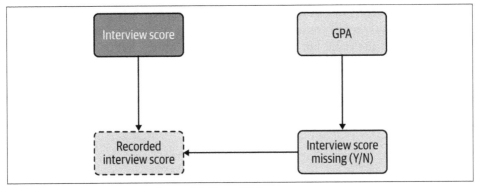

*Figure 6-19. Interview score being MAR on GPA*

 Rubin's classification of MCAR/MAR/MNAR is based solely on what the cause of missingness is. It doesn't take into account whether that causal relationship shows randomness or not. Here, counterintuitively, the fact that the missingness of *InterviewScore* is based deterministically on *GPA* makes *InterviewScore MAR* on *GPA* even though there's no randomness involved.

This can also happen for variables that are MNAR, where only values above or below a certain threshold get recorded. For instance, only values outside of a normal range may be saved in a file, or only people under or above a certain threshold will register themselves (e.g., for tax purposes).

In between these two extremes of complete randomness and complete determinism (either of the MAR or MNAR type), there are situations where the probability of missingness increases or decreases continuously based on the values of the cause of missingness.

Figure 6-20 shows what this looks like in the simplest case of two variables, X and Y, where X has missing values. For the sake of readability, available values are shown as full squares, whereas "missing" values are shown as crosses. The first row of Figure 6-20 shows scatterplots of Y against X and the second row shows for each one of these a line plot of the relationship between X and the probability of missingness:

- The leftmost column shows X being MCAR. The probability of missingness is constant at 0.5 and independent of X. Squares and crosses are spread similarly across the plot.

- The central columns show X being probabilistically MNAR with increasing strength. Squares are more common on the left of the plot and crosses more common on the right, but there are still crosses on the left and squares on the right.

- The rightmost column shows X being deterministically MNAR. All values of X below 5 are available (squares) and all the values above 5 are "missing" (crosses).

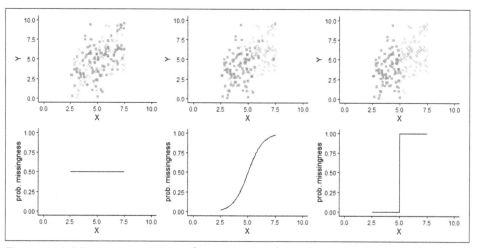

*Figure 6-20. Missingness spectrum, from MCAR (leftmost) to deterministic MNAR (rightmost) through probabilistic MNAR (center)*

This spectrum of missingness is rarely discussed in statistical treatments of missing data, because it is difficult to confirm through purely mathematical methods. But this is a book about behavioral analytics, so we can and should use common sense and business knowledge. In the GPA example, the threshold in data results from the application of a business rule that you should be aware of. In most situations, you expect a variable to be in a certain range of values, and you should have a sense of how likely it is for a possible value to not be represented in your data.

In our AirCnC survey data, we have three personality traits: *Openness*, *Extraversion*, and *Neuroticism*. In real life, these variables would result from the aggregation of the answers to several questions and would have a bell-shaped distribution over a known interval (see Funder (2016) for a good introduction to personality psychology). Let's assume that the relevant interval is 0 to 10 in our data and let's look at the distribution of our variables (Figure 6-21).

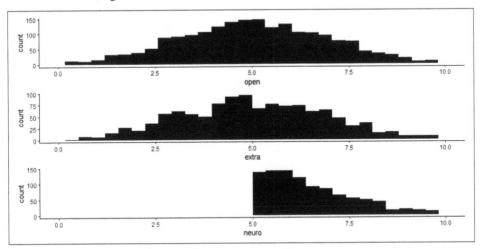

*Figure 6-21. Histograms of personality traits in our data*

Clearly, something is going on with *Neuroticism*. Based on how the personality traits are constructed, we would expect the same type of curve for all three variables, and we would certainly not expect to have a large number of customers with a value of 5, and none with a value of 4. This overwhelmingly suggests a deterministically MNAR variable, which we'll have to handle accordingly.

You should now be able to form a reasonable opinion of the pattern of missingness in a data set. How many missing values are there? Does their missingness appear related to the values of the variable itself (MNAR), another variable (MAR), or neither (MCAR)? Are these missingness relationships probabilistic or deterministic?

Figure 6-22 presents a decision tree recapping our logic to diagnose missing data.

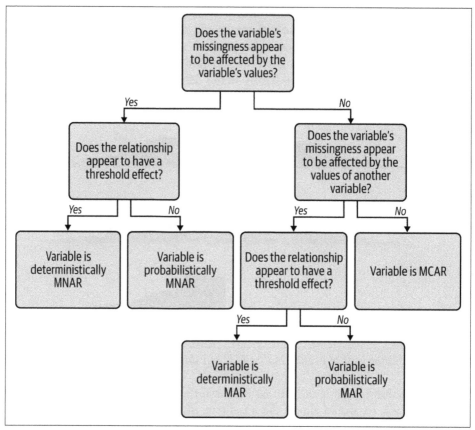

*Figure 6-22. Decision tree to diagnose missing data*

In the next section, we'll see how to deal with missing data in each of these cases.

# Handling Missing Data

The first thing to keep in mind as we get into the how-to section of the chapter is that we're not trying to deal with missing data for the sake of it: our goal is to obtain unbiased and accurate estimates of causal relationships in our data. Missing data is problematic only to the extent that it interferes with that goal.

This bears emphasizing, because your first instinct might be that the outcome of successfully addressing missing data is a data set with no missing data, and that's just not the case. The method we'll be using, multiple imputation (MI), creates multiple copies of your data, each one with its own imputed values. In layman's terms, we will never be saying "the correct replacement for Bob's missing age is 42" but rather "Bob might be 42, 38, 44, 42, or 38 years old." There is no single best guess, but instead a distribution of possibilities. Another best-practice approach, maximum likelihood estimation, doesn't even make any guess for individual values and only deals with higher-order coefficients such as means and covariances.

In the next subsection, I will give you a high-level overview of the MI approach. After that, we'll dive into more detailed algorithm specifications for the model:

1. First, the predictive mean matching algorithm
2. Then, the normal algorithm
3. Finally, how to add auxiliary variables to an algorithm

Unfortunately, there isn't a one-to-one relationship between the type of missingness in Rubin's classifications and the appropriate algorithm specification, as the amount of information available also matters (Table 6-4).

*Table 6-4. Optimal MI parameters based on type of missingness and information available*

| Missingness type | No info | Distribution of variable is normal | Distribution of missingness is deterministic |
|---|---|---|---|
| MCAR | Mean matching | Normal | (not possible) |
| MAR | Mean matching | Normal | Normal + auxiliary variables |
| MNAR | Mean matching + auxiliary variables | Normal + auxiliary variables | Normal + auxiliary variables |

## Introduction to Multiple Imputation (MI)

To understand how multiple imputation works, it helps to contrast it with the traditional approaches to missing data. Apart from simply dropping all rows with missing values, traditional approaches all rely on replacing missing values with a specific value. The replacement value may be the overall mean of the variable, or the predicted values based on the other variables available for that customer. Regardless of the rule used for the replacement value, these approaches are fundamentally flawed because they ignore the additional uncertainty introduced by the presence of missing data, and they may introduce bias in our analyses.

The MI solution to this problem is, as its name indicates, to build multiple data sets where missing values are replaced by different values, then run our analysis of interest with each one of them, and finally aggregate the resulting coefficients.

In both R and Python, this whole process is managed behind the scenes, and if you want to keep it simple, you can just specify the data and analyses you want to run.

Let's look first at the R code:

```
## R
> MI_data <- mice(available_data, print = FALSE)
> MI_summ <- MI_data %>%
    with(lm(bkg_amt~age+open+extra+neuro+gender+state)) %>%
```

```
    pool() %>%
    summary()
> print(MI_summ)
          term    estimate  std.error  statistic       df      p.value
1 (Intercept) 240.990671 15.9971117 15.064636  22.51173 3.033129e-13
2         age  -1.051678  0.2267569 -4.637912  11.61047 6.238993e-04
3        open   3.131074  0.8811587  3.553360 140.26375 5.186727e-04
4       extra  11.621288  1.2787856  9.087753  10.58035 2.531137e-06
5       neuro  -6.799830  1.9339658 -3.516003  15.73106 2.929145e-03
6     genderF -11.409747  4.2044368 -2.713740  20.73345 1.310002e-02
7      stateB  -9.063281  4.0018260 -2.264786 432.54286 2.401986e-02
8      stateC  -5.334055  4.7478347 -1.123471  42.72826 2.675102e-01
```

The `mice` package (Multiple Imputation by Chained Equations) has the `mice()` function, which generates the multiple data sets. We then apply our regression of interest to each one of them by using the keyword `with()`. Finally, the `pool()` function from mice aggregates the results in a format that we can read with the traditional `summary()` function.

The Python code is almost identical, because it implements the same approach:

```python
## Python
MI_data_df = mice.MICEData(available_data_df)
fit = mice.MICE(model_formula='bkg_amt~age+open+extra+neuro+gender+state',
                model_class=sm.OLS, data=MI_data_df)
MI_summ = fit.fit().summary()
print(MI_summ)
```

```
                         Results: MICE
=====================================================================
Method:            MICE          Sample size:      2000
Model:             OLS           Scale             5017.30
Dependent variable: bkg_amt      Num. imputations  20
---------------------------------------------------------------------
          Coef.    Std.Err.    t     P>|t|   [0.025   0.975]   FMI
---------------------------------------------------------------------
Intercept 120.3570  8.8662 13.5748 0.0000 102.9795 137.7344 0.4712
age        -1.1318  0.1726 -6.5555 0.0000  -1.4702  -0.7934 0.2689
open        3.1316  0.8923  3.5098 0.0004   1.3828   4.8804 0.1723
extra      11.1265  1.0238 10.8680 0.0000   9.1200  13.1331 0.3855
neuro      -4.5894  1.7968 -2.5542 0.0106  -8.1111  -1.0677 0.4219
gender_M   65.9603  4.8191 13.6873 0.0000  56.5151  75.4055 0.4397
gender_F   54.3966  4.6824 11.6171 0.0000  45.2192  63.5741 0.4154
state_A    40.9352  3.9080 10.4748 0.0000  33.2757  48.5946 0.3921
state_B    37.3490  4.0727  9.1706 0.0000  29.3666  45.3313 0.2904
state_C    42.0728  3.8643 10.8875 0.0000  34.4989  49.6468 0.2298
=====================================================================
```

Here, the `mice` algorithm is imported from the `statsmodels.imputation` package. The `MICEData()` function generates the multiple data sets. We then indicate through the `MICE()` function the model formula, regression type (here, `statsmodels.OLS`), and data we want to use. We fit our model with the `.fit()` and `.summary()` methods before printing the outcome.

 One complication with the Python implementation of mice is that it doesn't accommodate categorical variables as predictors. If you really want to use Python nonetheless, you'll have to one-hot encode categorical variables first. The following code shows how to do it for the *Gender* variable:

```Python
## Python
gender_dummies = pd.get_dummies(available_data_df.\
                                gender,
                                prefix='gender')
available_data_df = pd.concat([available_data_df,
                               gender_dummies],
                               axis=1)
available_data_df.gender_F = \
np.where(available_data_df.gender.isna(),
         float('NaN'), available_data_df.gender_F)
available_data_df.gender_M = \
np.where(available_data_df.gender.isna(),
         float('NaN'), available_data_df.gender_M)
available_data_df = available_data_df.\
drop(['gender'], axis=1)
```

First, we use the `get_dummies()` function from pandas to create variables `gender_F` and `gender_M`. After adding these columns to our dataframe, we indicate where the missing values are (by default, the one-hot encoding function sets all binary variables to 0 when the value for the categorical variable is missing). Finally, we drop our original categorical variable from our data and fit our model with the new variables included.

However, one-hot encoding breaks some of the internal structure of the data by removing the logical connections between variables, so your mileage may vary (e.g., you can see that the coefficients for the categorical variables are different between R and Python because of the different structures) and I would encourage you to use R instead if categorical variables play an important role in your data.

Voilà! If you were to stop reading this chapter right now, you would have a solution to handle missing data that would be significantly better than the traditional approaches. However, we can do even better by taking the time to lift the hood and better understand the imputation algorithms.

# Default Imputation Method: Predictive Mean Matching

In the previous subsection, we left the imputation method unspecified and relied on mice's defaults. In Python, the only imputation method available is predictive mean matching, so there's nothing to do there. Let's check what the default imputation methods are in R by asking for a summary of the imputation process:

```
## R
> summary(MI_data)
Class: mids
Number of multiple imputations:  5
Imputation methods:
       age      open     extra    neuro    gender     state   bkg_amt
        ""        ""     "pmm"    "pmm"        ""  "logreg"     "pmm"
PredictorMatrix:
       age open extra neuro gender state bkg_amt
age      0    1     1     1      1     1       1
open     1    0     1     1      1     1       1
extra    1    1     0     1      1     1       1
neuro    1    1     1     0      1     1       1
gender   1    1     1     1      0     1       1
state    1    1     1     1      1     0       1
```

That's a lot of information. For now, let's look only at the Imputation methods line. Variables that don't have missing data have an empty field "", which makes sense because they don't get imputed. Categorical variables have the method logreg, i.e., logistic regression. Finally, numeric variables have the method pmm, which stands for predictive mean matching (PMM). The pmm method works by selecting the closest neighbors of the individual with the missing value and replacing the missing value with the value of one of the neighbors. Imagine, for example, a data set with only two variables: *Age* and *ZipCode*. If you have a customer from zip code 60612 with a missing age, the algorithm will pick at random the age of another customer in the same zip code, or as close as possible.

Because of some randomness baked in the process, each of the imputed data sets will end up with slightly different values, as we can visualize through the convenient den sityplot() function from the mice package in R:

```
## R
> densityplot(MI_data, thicker = 3, lty = c(1,rep(2,5)))
```

Figure 6-23 shows the distributions of the numeric variables in the original available data (thick line) and in the imputed data sets (thin dashed lines). As you can see, the distributions stick pretty close to the original data; the exception is *BookingAmount*, which is overall more concentrated around the mean (i.e., "higher peaks") in the imputed data set than in the original data.

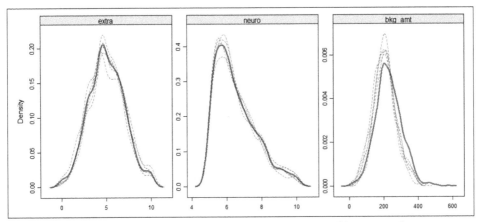

*Figure 6-23. Distributions of imputed values for numeric variables in our data*

PMM has some important properties that may or may not be desirable, depending on the context. The most important property is that it's basically an interpolation method. Therefore, you can picture PMM as creating values that are between existing values. By doing so, it minimizes the risk of creating nonsensical situations such as pregnant fathers or negative amounts. This approach will also work well when a variable has a weird distribution, such as *Age* in our data, which has two peaks, because it makes no assumptions about the shape of the overall distribution; it just grabs a neighbor.

There are several downsides to PMM, however: it's slow and doesn't scale well to large data sets, because it must constantly recalculate distances between individuals. In addition, when you have many variables or a lot of missing values, the closest neighbors may be "far away," and the quality of the imputation will deteriorate. This is why PMM will not be our preferred option when we have distributional information, as we'll see in the next subsection.

## From PMM to Normal Imputation (R Only)

While PMM is a decent starting point, we often have information about the distribution of numeric variables that we can leverage to speed up and improve our imputation models in R. In particular, behavioral and natural sciences often assume that numeric variables follow a normal distribution, because it is very common. When that's the case, we can fit a normal distribution to a variable and then draw imputation values from that distribution instead of using PMM. This is done by creating a vector of imputation methods, with the value `"norm.nob"` for the variables for which we'll assume normality, and then passing that vector to the `parameter` method of the `mice()` function:

```
## R
> imp_meth_dist <- c("pmm", rep("norm.nob",3), "", "logreg", "norm.nob")
> MI_data_dist <- mice(available_data, print = FALSE, method = imp_meth_dist)
```

As you can see, the syntax is very simple. The only question is to determine for which of the numeric variables we want to use a normal imputation. Let's look at the numeric variables in our available data (Figure 6-24).

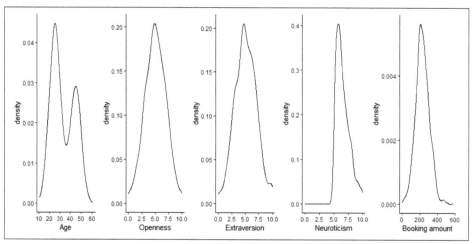

*Figure 6-24. Distribution of numeric variables in our data*

*Age* is obviously not normal with its two peaks, but all the other variables have only one peak. *Openness, Extraversion,* and *BookingAmount* also appear reasonably symmetrical (in technical terms, they're not skewed). Statistical simulations show that as long as a variable is one-peaked and doesn't have a "fat tail" in only one direction, assuming normality does not introduce bias. Therefore, we can assume normality for *Openness, Extraversion,* and *BookingAmount*.

As we saw in the previous section, *Neuroticism* presents an unusual asymmetrical pattern: values are restricted to [5,10] even though the psychological scale we're using goes from 0 to 10, which suggests that *Neuroticism* might be "deterministically" MNAR, i.e., all values of *Neuroticism* below a certain threshold are missing. Using PMM for imputation is problematic in such a situation: there are only a few or no neighbors for imputation on a significant range of values. At the extreme, all the missing values of X would be imputed as 5, the value of the threshold. This is a situation where the normal method will be better able to recover the true missing values. We can see that by comparing the values imputed by the two methods. Figure 6-25 shows the available values of *Neuroticism* (squares) and the values imputed with the PMM method (crosses, top panel) and with the normal method (crosses, bottom panel).

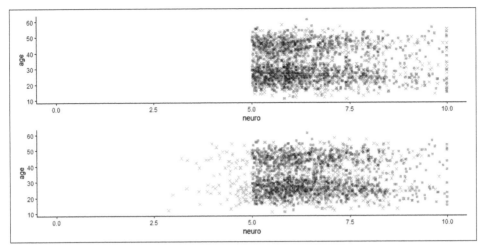

*Figure 6-25. PMM imputation (top) and normal imputation (bottom) with a deterministically MNAR variable*

As you can see, the PMM method doesn't impute any value for *Neuroticism* below 5, whereas the normal method does. In addition, the PMM method imputes too many values close to 10, whereas the normal method more adequately captures the overall shape of the distribution. Still, the normal method is far from recovering the true distribution (which goes all the way to zero). This is a common problem for variables that are deterministically MAR or MNAR. We can further improve on the normal imputation by using auxiliary variables, as we'll see in the next subsection.

## Robust Imputation

If you have variables that are one-peaked but not normal (e.g., at least one tail of the distribution is unusually fat or thin compared to the normal distribution) and you care about the extra accuracy, the R package ImputeRobust builds on the mice package to refine your imputation model. See the CRAN vignette (*https://oreil.ly/5quMo*) for more details.

## Adding Auxiliary Variables

Quite often, we'll have variables that are correlated with one of our variables with missing data (e.g., causes or effects of that variable) but don't belong in our regression model. This is a situation where the mice algorithm especially shines, because we can add these variables to our imputation model to increase its accuracy. They are then referred to as the "auxiliary variables" of our imputation model.

For our AirCnC example, the supplementary available data set contains two variables, *Insurance* and *Active*. The former indicates the amount of travel insurance bought by the customer and is strongly correlated with *Neuroticism*, while the latter measures the degree to which the customer has picked active vacations (e.g., rock climbing) and is strongly correlated with *Extraversion*. We'll use them to help impute the two personality variables.

Adding auxiliary variables to the imputation model is extremely simple: we simply need to add them to our data set before the imputation phase:

```R
## R
augmented_data <- cbind(available_data, available_data_supp)
MI_data_aux <- mice(augmented_data, print = FALSE)
```

```Python
## Python
augmented_data_df = pd.concat([available_data_df, available_data_supp_df],
                             axis=1)
MI_data_aux_df = mice.MICEData(augmented_data_df)
```

We can then run all our analyses as before. When adding auxiliary variables, it generally makes sense to use the normal method for the variables correlated with our auxiliary variables (here *Neuroticism* and *Extraversion*), especially when these variables are truncated or MNAR.

Apart from computation constraints, there are no limits to the number of auxiliary variables we can include. However, a potential risk is that some of our auxiliary variables may misleadingly appear correlated with a variable in our original data set just out of sheer randomness, e.g., *Insurance* appearing correlated with *Extraversion* even though it isn't truly. Such a "false positive" correlation would then be unduly reinforced by the imputation model.

The solution to that potential problem is to restrict auxiliary variables to be used only for the imputation of certain variables. Unfortunately, this solution is available only in R. This is where the predictor matrix of the mice() function comes in. This matrix appears when printing the summary of the imputation phase, and can also be extracted directly from our MIDS object:

```R
## R
> pred_mat <- MI_data_aux$predictorMatrix
> pred_mat
          age open extra neuro gender state bkg_amt insurance active
age         0    1     1     1      1     1       1         1      1
open        1    0     1     1      1     1       1         1      1
extra       1    1     0     1      1     1       1         1      1
neuro       1    1     1     0      1     1       1         1      1
gender      1    1     1     1      0     1       1         1      1
state       1    1     1     1      1     0       1         1      1
bkg_amt     1    1     1     1      1     1       0         1      1
insurance   1    1     1     1      1     1       1         0      1
active      1    1     1     1      1     1       1         1      0
```

This matrix indicates which variable is used to impute which. By default, all variables are used to impute all variables except themselves. A "1" in the matrix indicates that the "column" variable is used to impute the "row" variable. We'll therefore want to modify the last two columns, for *Insurance* and *Active*, so that they'll be used only to impute *Neuroticism* and *Extraversion* respectively:

```R
## R
> pred_mat[,"insurance"] <- 0
> pred_mat[,"active"] <- 0
> pred_mat["neuro","insurance"] <- 1
> pred_mat["extra","active"] <- 1
> pred_mat
          age open extra neuro gender state bkg_amt insurance active
age         0    1     1     1      1     1       1         0      0
open        1    0     1     1      1     1       1         0      0
extra       1    1     0     1      1     1       1         0      1
neuro       1    1     1     0      1     1       1         1      0
gender      1    1     1     1      0     1       1         0      0
state       1    1     1     1      1     0       1         0      0
bkg_amt     1    1     1     1      1     1       0         0      0
insurance   1    1     1     1      1     1       1         0      0
active      1    1     1     1      1     1       1         0      0
```

With that modification, we'll reduce the risk of inadvertently baking in fluke correlations into our imputation model.

## Scaling Up the Number of Imputed Data Sets

The default number of imputed data sets created by the mice algorithm is 5 in R and 10 in Python. These are fine defaults for exploratory analyses.

For your final run, you should use 20 (by passing m=20 as a parameter setting to the mice() function) if you're interested only in the estimated values of the regression coefficients. If you want more precise information such as confidence intervals or interactions between variables, you might want to aim for 50 to 100. The main constraints then become the computer's speed and memory—if your data set is 100 Mb or even 1 Gb do you have the RAM to create a hundred copies of it?—as well as your patience.

The syntax to change the number of imputed data sets is straightforward. In R it is passed as a parameter to the mice() function, while in Python it is passed as a parameter for the .fit() method of the MICE object:

```R
## R
MI_data <- mice(available_data, print = FALSE, m=20)
```

```python
## Python
fit = mice.MICE(model_formula='bkg_amt~age+open+extra+neuro+gender+state',
                model_class=sm.OLS, data=MI_data_df)
MI_summ = fit.fit(n_imputations=20).summary()
```

# Conclusion

Missing data can present a real problem in behavioral data analysis, but it doesn't have to. At a minimum, using the `mice` package in R or Python with its default parameters will outperform deleting all rows with missing values. By properly diagnosing missingness based on Rubin's classification, and leveraging all available information, you can generally do better than that. To recap the decision rules in one place, Figure 6-26 shows the decision tree to diagnose missing data and Table 6-5 the optimal MI parameters based on type of missingness and information available.

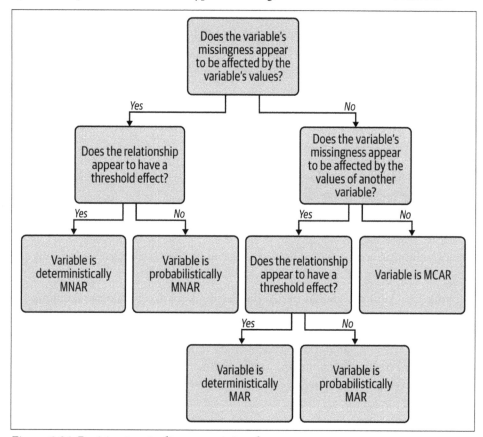

*Figure 6-26. Decision tree to diagnose missing data*

*Table 6-5. Optimal MI parameters based on type of missingness and information available*

| Type of missingness | No info | Variable distribution is normal | Missingness distribution is deterministic |
|---|---|---|---|
| MCAR | PMM | norm.nob | |
| MAR | PMM | norm.nob | norm.nob + aux. var. |
| MNAR | PMM + aux. var. | norm.nob + aux. var. | norm.nob + aux. var. |

# CHAPTER 7

# Measuring Uncertainty with the Bootstrap

With ideal data, you are now able to draw robust conclusions from behavioral data and measure the causal impact of a business/environment change on human behaviors. But how can you proceed if you have suboptimal data? In academic research, one can always fall back to the null hypothesis when faced with inconclusive data and refuse to pass judgment. But in applied research there is no null hypothesis, only alternative courses of action to choose from.

Small sample sizes, weirdly shaped variables, or situations that require advanced analytical tools (e.g., hierarchical modeling, which we'll see later in the book) can all result in shaky conclusions. Certainly, a linear regression algorithm will spit out a coefficient under all but the most extreme cases, but should you trust it? Can you confidently advise your boss to stake millions of dollars on it?

In this chapter, I'll introduce you to an extremely powerful and general simulation tool, the Bootstrap, which will allow us to draw robust conclusions from any data, however small or weird. It works by creating and analyzing slightly different versions of your data based on random numbers. A great feature of the Bootstrap is that you literally can never go wrong by applying it: in situations that are best-case scenarios for traditional statistical methods (e.g., running a basic linear regression on a large and well-behaved data set), the Bootstrap is slower and less accurate, but it is still in the ballpark. But as soon as you move away from such best-case scenarios, the Bootstrap quickly outperforms traditional statistical methods, often by a wide margin.[1] Therefore, we'll rely on it extensively throughout the rest of the book. In particular, we'll use it when designing and analyzing experiments in Part IV, to build simulated equivalents of p-values that are more intuitive than the traditional, statistical, ones.

---

1 See Wilcox (2010), which shows the danger of assuming normality as a matter of course.

In the first section, we'll focus on exploratory/descriptive data analysis, and we'll see that the Bootstrap can already be of use at that stage. In the second section, we'll use the Bootstrap in the context of regression. We'll then broaden our perspective to discuss when to use the Bootstrap and what tools you can use to make your life easier with it.

# Intro to the Bootstrap: "Polling" Oneself Up

While our ultimate goal is to use the Bootstrap for regression, we can start with the simpler example of descriptive statistics: getting the mean of a sample data set.

## Packages

Running Bootstrap simulations is one situation where Python gets a head start, thanks to the relatively good performance of its loops. We'll use the following packages for diagnostic, but otherwise raw Python is all you need in most cases:

```
## Python
import statsmodels.api as sm # For QQ-plot
import statsmodels.stats.outliers_influence as st_inf # For Cook distance
```

But don't worry, R users. I'll show you how the Bootstrap works with "for" loops, so that you can understand its logic, and at the end of the chapter I'll give you two higher-performance alternatives, including the package for this book.

## The Business Problem: Small Data with an Outlier

C-Mart's management is interested in understanding how long it takes its bakers to prepare made-to-order cakes, for the purpose of possibly revising its pricing structure. To that end, they have asked C-Mart's industrial engineer to do a time study. As its name indicates, a time study (a.k.a. time-and-motion study) is the direct observation of a production process to measure the duration of the tasks involved. Given that the process is time-consuming (pun intended), the engineer has selected ten different stores that are somewhat representative of C-Mart's business. In each store they observed one baker preparing one cake. They also recorded each baker's work experience, measured in months on the job.

All together the engineer has 10 observations, which is not a very large sample size to begin with. Even if all of the data conformed very consistently to a clear relationship, the sample size alone would suggest using the Bootstrap. However, when exploring their data, the engineer observed the presence of an outlier (Figure 7-1).

We have one extreme point in the upper left corner, corresponding to a new employee who spent most of a day on a complex cake for a corporate retreat. How should the engineer report the data from their study? They might be tempted to treat the largest observation as an outlier, which is the polite way of saying "discard it and pretend it didn't happen." But that observation, while unusual, is not an aberration per se. There was no measurement error, and those circumstances probably occur from time to time. An alternative would be to only report the overall mean duration, 56 minutes, but that would also be misleading because it would not convey the variability and uncertainty in the data. The traditional recommendation in that situation would be to use a confidence interval around the mean. Let's calculate the normal 95% confidence interval through a regression. (Using a regression is overkill in this case—there are much simpler ways to calculate a mean—but it will serve as a gentle introduction to the process we'll use later in the chapter.)

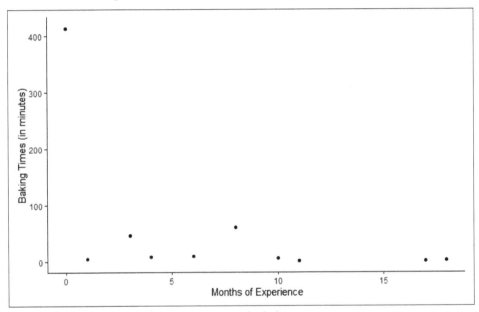

*Figure 7-1. Experience and preparation time by baker*

We first run the regression `times~1`, i.e., with only the intercept. We then extract the resulting estimate for the intercept coefficient that, in case you're not familiar with that calculation, is equal to the mean of our dependent variable. We also extract the standard error for that coefficient. As you'll learn in any stats class, the lower limit of a normal 95%-CI is equal to the mean minus 1.96 times the standard error, and the upper limit is equal to the mean plus 1.96 times the standard error:

```
## R (output not shown)
lin_mod_summ <- summary(lm(times~1, data=dat))
est <- lin_mod_summ$coefficients[1,1]
se <- lin_mod_summ$coefficients[1,2]
LL <- est-1.96*se
UL <- est+1.96*se
```

```
## Python
lin_mod = ols("times~1", data=data_df).fit()
est = lin_mod.params['Intercept']
se = lin_mod.bse['Intercept']
LL = est-1.96*se #Lower limit
UL = est+1.96*se #Upper limit
print("LL = ", LL)
print("UL = ",UL)

LL =  -23.040199740431333
UL =  134.64019974043134
```

Unfortunately, the 95%-CI in this case is [−23; 135], which is obviously nonsensical because duration times can't be negative. This happened because traditional CIs assume that the variable at hand follows a normal distribution around its mean, which in this case is incorrect. We can imagine that the engineer's audience would not take too kindly to negative durations, but that is one of the problems that the Bootstrap can solve.

## Bootstrap Confidence Interval for the Sample Mean

The Bootstrap allows us to make full use of the data that we do have available and to draw reasonable conclusions regardless of sample size or data shape challenges. It does so by creating multiple imaginary data sets based on the data that we have available. Comparing these data sets with each other allows us to cut through noise and more accurately represent the importance of outlier values. It can also provide tighter confidence intervals, since it removes some of the uncertainty created by noise.

This is different from just choosing a narrower range from the start (e.g., selecting the 80%-CI instead of the 95%-CI) because the Bootstrap's generated data sets reflect true probability distributions given the available data. There will be no generated data set with a negative duration because the data does not reflect that possibility, but there will be data sets that reflect very long durations because the original data does include that as a possibility. So a confidence interval generated using the Bootstrap would be expected to remove more from the negative side of the range, but it might not remove as much (or might even add to) the positive side of the range.

The process to build a Bootstrap CI is conceptually simple:

1. We simulate new samples of the same size by drawing with replacement from our observed sample.

2. Then, for each simulated sample we calculate our statistic of interest (here the mean, which is what our industrial engineer wants to measure).

3. Finally, we build our CI by looking at the percentiles of the values obtained in step 2.

Drawing with replacement means that each value has the same probability of being drawn each time, regardless of whether or not it has already been drawn.

For example, drawing with replacement from (A, B, C) is equally likely to yield (B, C, C) or (A, C, B) or (B, B, B), etc. Because there are three possibilities for each of the three positions, there are 3 x 3 x 3 = 27 possible simulated samples. If we drew without replacement, this would mean that a value cannot be drawn more than once, and the only possible combinations would be permutations of the original sample, such as (A, C, B) or (B, A, C). This would simply amount to shuffling the values around, which would be pointless because the mean (or any other statistic of interest) would remain exactly the same.

Drawing with replacement is very simple in both R and Python:

```
## R
boot_dat <- slice_sample(dat, n=nrow(dat), replace = TRUE)

## Python
boot_df = data_df.sample(len(data_df), replace = True)
```

## Bootstrap and the Meaning of Statistics

Why are we drawing with replacement? To really grasp what's happening with the Bootstrap, it's worth taking a step back and remembering the point of statistics: we have a population that we cannot fully examine, so we're trying to make inferences about this population based on a limited sample. To do so, we create an "imaginary" population through either statistical assumptions or the Bootstrap. With statistical assumptions we say, "imagine that this sample is drawn from a population with a normal distribution," and then make inferences based on that. With the Bootstrap we're saying, "imagine that the population has exactly the same probability distribution as the sample," or equivalently, "imagine that the sample is drawn from a population made of a large (or infinite) number of copies of that sample." Then drawing with replacement from that sample is equivalent to drawing without replacement from that imaginary population. As statisticians will say, "the Bootstrap sample is to the sample what the sample is to the population."

The beauty of generating new samples by drawing only from our observed sample is that it avoids making any distributional assumption about data outside of the sample we observed. To see what this means, let's simulate B = 2,000 Bootstrap samples (to avoid confusion, I'll always use B for the number of Bootstrap samples and N for the sample size) and calculate the mean of each. Our code proceeds as follows (the callout numbers are shared between R and Python):

```
## R
mean_lst <- list()  ❶
B <- 2000
```

```
N <- nrow(dat)
for(i in 1:B){ ❷
  boot_dat <- slice_sample(dat, n=N, replace = TRUE)
  M <- mean(boot_dat$times)
  mean_lst[[i]] <- M}
mean_summ <- tibble(means = unlist(mean_lst)) ❸

## Python
res_boot_sim = [] ❶
B = 2000
N = len(data_df)
for i in range(B): ❷
    boot_df = data_df.sample(N, replace = True)
    M = np.mean(boot_df.times)
    res_boot_sim.append(M)
```

❶ First I initialize an empty list for the results, as well as B and N.

❷ Then I use a for loop to generate the Bootstrap samples by drawing with replacement from the original data, each time calculating the mean and adding it to the list of results.

❸ Finally, in R, I reformat the list into a tibble, for ease of use with ggplot2.

Figure 7-2 shows the distribution of the means.

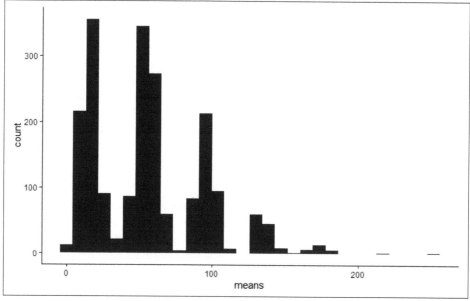

*Figure 7-2. Distribution of the means of 2,000 samples*

As you can see, the histogram is very irregular: there is a big peak close to the mean of our original data set along with smaller peaks corresponding to certain patterns. Given how extreme our outlier is, each of the seven peaks corresponds to its number of repetitions in the Bootstrap sample, from zero to six. In other words, it doesn't appear at all in the samples whose means are in the first (leftmost) peak, it appears exactly once in the samples whose means are in the second peak, and so on. It's worth noting that even if we increased the number of Bootstrap samples, the irregularity of the histogram would not disappear (i.e., the "valleys" between the peaks would not get filled), because it reflects the roughness of our data and not the limitations of our random process. The range of values within our data is so extreme that the highest possible means when the outlier is excluded are still rarely high enough to meet the lowest possible means when the outlier is included. If the value of the outlier were cut in half, and thus were closer to the rest of the population, the histogram would appear to smooth out considerably because the edges of the outlier count peaks would overlap each other.

The number of Bootstrap samples matters nonetheless, but for a different reason: the higher that number, the more you'll be able to see very unlikely samples and therefore extreme values. Here, the absolute highest possible value for a sample mean would be 413 if we drew the outlier 10 times. This has a probability of $(0.1)^{10}$ (one-tenth to the power of 10), meaning it will happen about one time per 10 billion samples. With our mere 2,000 samples, we are barely seeing values around 200. But the overall mean or median of our samples would remain the same plus or minus negligible sampling variations.

Here are some general guidelines for number of samples:

- 100 to 200 samples to get an accurate central estimate (e.g., a coefficient in a regression; it's called "central" because it's roughly speaking at the center of a CI, as opposed to the CI's bounds or limits)
- 1,000 to 2,000 samples to get accurate 90%-CI bounds
- 5,000 samples to get accurate 99%-CI bounds

In general, start low, and if in doubt increase the number and try again. This is fundamentally different from, for example, running multiple analyses on your data until you get numbers you like (a.k.a. "p-value hacking" or "p-hacking"); it's more like changing the resolution of your screen when looking at a figure. It entails no risk for your analyses; it simply takes more or less of your time, depending on the size of your data and the computational power of your machine.

Given the data that we have, the only way we could increase the smoothness of the histogram would be to increase the sample size. However, we would have to increase the size of the original sample from the real world, not the size of the Bootstrap samples. Why can't we increase the size of the Bootstrap samples (e.g., drawing 100 values with replacement from our sample of 10 values)? Because our goal is not to create new samples but to determine how far off our estimate of the mean could be when we make the assumption that the population is proportionally identical to our original sample. To do this we need to use all the information in the original sample—no less and no more. Creating larger samples from our 10 original values would be "pretending" that we have more information than we actually have.

The engineer is ready to use the Bootstrap to determine the bounds of the CI for the duration of cake preparation. These bounds are determined from the *empirical* distribution of the preceding means. This means that instead of trying to fit a statistical distribution (e.g., normal), they can simply order the values from smallest to largest and then look at the 2.5% quantile and the 97.5% quantile to find the two-tailed 95%-CI. With 2,000 samples, the 2.5% quantile is equal to the value of the 50th smallest mean (because 2,000 * 0.025 = 50), and the 97.5% quantile is equal to the value of the 1950th mean from smaller to larger, or the 50th largest mean (because both tails have the same number of values). Fortunately, we don't have to calculate these by hand:

```
## R (output not shown)
LL_b <- as.numeric(quantile(mean_summ$means, c(0.025)))
UL_b <- as.numeric(quantile(mean_summ$means, c(0.975)))

## Python
LL_b = np.quantile(res_boot_sim, 0.025)
UL_b = np.quantile(res_boot_sim, 0.975)
print("LL_b = ", LL_b)
print("UL_b = ",UL_b)

# Note that the following values are random and will vary across draws
LL_b =  5.90
UL_b =  137.9025
```

The Bootstrap 95%-CI is [5.90; 137.90] (plus or minus some sampling difference), which is much more realistic. Figure 7-3 shows the same histogram as Figure 7-2 but adds the mean of the means, the normal CI bounds and the Bootstrap CI bounds.

In addition to the Bootstrap lower bound being above zero, we can also note that the Bootstrap upper bound is slightly higher than the normal upper bound, which better reflects the asymmetry of the distribution toward the right.

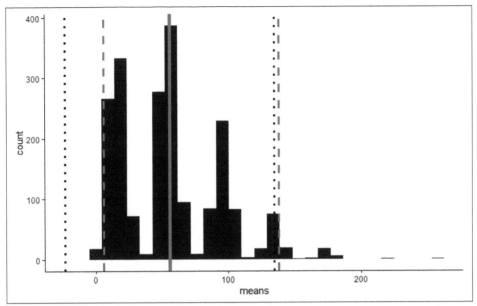

*Figure 7-3. Distribution of the means of 2,000 samples, with mean of the means (thick line), normal 95%-CI bounds (dotted lines), and Bootstrap CI bounds (dashed lines)*

## Bootstrap Confidence Intervals for Ad Hoc Statistics

Using the Bootstrap has allowed us to build a reasonable CI when the traditional statistical approach was failing. We can also use it to build CIs in situations where there is no other way to do so. Let's imagine, for example, that C-Mart's management is considering instituting a time promise—"your cake in three hours or 50% off"—and wants to know how often a cake currently takes more than three hours to be baked. Our estimate would be the sample percentage: it happens in 1 of the 10 observed cases, or 10%. But we can't leave it at that, because there is significant uncertainty around that estimate, which we need to convey. Ten percent out of 10 observations is much more uncertain than 10% out of 100 or 1,000 observations.

So how can we build a CI around that 10% value? With the Bootstrap, of course. The process is exactly the same as earlier, except that instead of taking the mean of each simulated sample, we'll measure the percentage of the values in the sample that are above 180 minutes:

```
## R
promise_lst <- list()
N <- nrow(dat)
B <- 2000
for(i in 1:B){
  boot_dat <- slice_sample(dat, n=N, replace = TRUE)
  above180 <- sum(boot_dat$times >= 180)/N
```

```
    promise_lst[[i]] <- above180}
promise_summ <- tibble(above180 = unlist(promise_lst))
LL_b <- as.numeric(quantile(promise_summ$above180, c(0.025)))
UL_b <- as.numeric(quantile(promise_summ$above180, c(0.975)))

## Python
promise_lst = []
B = 2000
N = len(data_df)
for i in range(B):
    boot_df = data_df.sample(N, replace = True)
    above180 =  len(boot_df[boot_df.times >= 180]) / N
    promise_lst.append(above180)
LL_b = np.quantile(promise_lst, 0.025)
UL_b = np.quantile(promise_lst, 0.975)
```

The histogram of the results is shown in Figure 7-4. There's "white space" between the bars again because we have only 10 data points, so percentages are multiples of 10%. That would not be the case with more data points; in general percentages will be multiples of 1/N with N the sample size (e.g., with 20 points, percentages would be multiples of 5%).

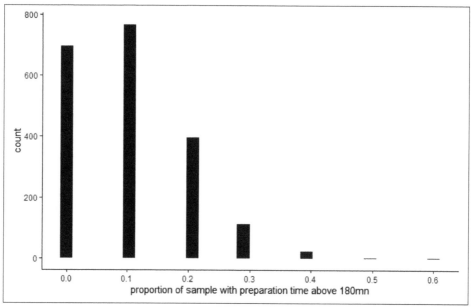

*Figure 7-4. Histogram of count of samples with a given proportion of durations above 180 minutes*

In about 700 of the 2,000 simulated samples, there was no cake with a preparation time above 180 minutes. In about 750, there was exactly one such cake, and so on. The corresponding 95%-CI is [0; 0.3]: the 50th lowest value is 0 and the 50th highest value is 0.3.

In other words, even with such limited data, we can quite confidently say that it's very unlikely (although not impossible) that more than 30% of the cakes take more than three hours to prepare. That's still a pretty large confidence interval, but not too shabby for only 10 observations and such a unique statistic!

 If this is hard to wrap your mind around, you can reframe the preceding problem by calculating the confidence interval for a binomial distribution with 1 success in 10 observations. Approximation methods are available in R and Python to calculate CIs in that case. These tend to be more conservative (i.e., broader) than our Bootstrap CI, but not vastly so.

By using the Bootstrap, the engineer can sharpen the analyses they would regularly want to perform with their data. They're able to use limited data to answer a variety of questions with a reasonable amount of certainty (and correspondingly tolerable uncertainty).

## The Bootstrap for Regression Analysis

While building a confidence interval around the mean can be useful, regression is really what this book is about, so let's see how we can use the Bootstrap for that purpose. Our industrial engineer at C-Mart wants to determine the effect of experience on baking time using the same data about cake preparation. The corresponding CD is very simple (Figure 7-5).

*Figure 7-5. Causal diagram for our relationship of interest*

Running a regression on our data is straightforward, given that the causal diagram did not reveal any confounder. However, the resulting coefficient is not significant:

```
## Python (output not shown)
print(ols("times~experience", data=data_df).fit().summary())

## R
mod <- lm(times~experience, data=dat)
mod_summ <- summary(mod)
mod_summ
...
Coefficients:
            Estimate Std. Error t value Pr(>|t|)
(Intercept) 132.389    61.750    2.144   0.0644
experience   -9.819     6.302   -1.558   0.1578
...
```

Our estimated coefficient is –9.8, meaning that every additional month of experience is expected to remove 9.8 minutes of preparation time. However, the traditional CI based on the regression standard error would be [–22.2; 2.5]. From a traditional perspective, this would be game over: the CI includes zero, meaning that months of experience could have a positive, negative, or zero effect on baking time, so we would decline to draw any substantive conclusion. Let's see instead what the Bootstrap tells us. The process is exactly the same as before: we simulate samples of 10 data points by drawing with replacement from our original sample a large number of times, then save the regression coefficient. Last time we used B = 2,000 samples. This time let's use B = 4,000, as it makes the corresponding histogram look smoother (Figure 7-6):

```R
## R (output not shown)
reg_fun <- function(dat, B){
  N <- nrow(dat)
  reg_lst <- list()
  for(i in 1:B){
    boot_dat <- slice_sample(dat, n=N, replace = TRUE)
    summ <- summary(lm(times~experience, data=boot_dat))
    coeff <- summ$coefficients['experience','Estimate']
    reg_lst[[i]] <- coeff}
  reg_summ <- tibble(coeff = unlist(reg_lst))
  return(reg_summ)}
reg_summ <- reg_fun(dat, B=4000)
```

```Python
## Python (output not shown)
reg_lst = []
B = 4000
N = len(data_df)
for i in range(B):
    boot_df = data_df.sample(N, replace = True)
    lin_mod = ols("times~experience", data=boot_df).fit()
    coeff = lin_mod.params['experience']
    reg_lst.append(coeff)
LL_b = np.quantile(reg_lst, 0.025)
UL_b = np.quantile(reg_lst, 0.975)
```

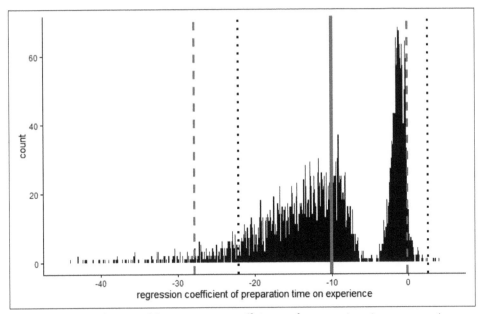

*Figure 7-6. Distribution of the regression coefficients of preparation time on experience, with their mean (thick line), Bootstrap CI bounds (thick dashed lines), and normal CI bounds (thin dotted lines) (B = 4,000 Bootstrap samples)*

The Bootstrap CI is [−28; −0.2]. As you can see in Figure 7-6, it's again asymmetric compared to the symmetric normal bounds, with a long tail to the left of the mean. The highly irregular shape of the distribution reflects the existence of two competing hypotheses:

- The tall and narrow peak near zero is made of samples that don't include the outlier, and as such it corresponds with the view that the outlier is a freak accident that won't repeat itself. That's the confidence interval you would get if you discarded the outlier.
- The broad and flat hump to the left is made of samples that include the outlier one or several times. It reflects the hypothesis that the outlier is truly representative of our data and that its true frequency may be even higher than in our small sample.

You can think of this as data-driven scenario analysis. What if this pattern didn't exist? What if it dominated our data? Instead of having to choose between discarding the outlier or letting it drive our results, the Bootstrap allows us to consider all possibilities at once.

In addition to building a CI, we can use the Bootstrap to determine the equivalent of a p-value. If you look at the output of our regression at the beginning of this section, you'll see the value 0.16 for experience in the column for the p-values (i.e., the column with the label Pr(>|t|)). You have probably already been told that a coefficient is statistically significant (i.e., statistically significantly different from zero) if its p-value is less than 0.05, or 0.01 in more stringent cases. Mathematically speaking, the p-value is such that the (1 minus p-value)-CI has zero as one of its bounds. In the case of the normal regression, zero is the upper bound of the 84%-CI. Because 84% is less than 95% or 99%, the coefficient for experience would not be considered statistically significant. The exact same logic can be used with the Bootstrap; we just have to calculate the fraction of the Bootstrap sample whose coefficient is above zero, and multiply it by 2 because it's a two-tailed test:[2]

```
## Python (output not shown)
pval = 2 * sum(1 for x in reg_lst if x > 0) / B

## R
reg_summ %>% summarise(pval = 2 * sum(coeff > 0)/n())
# A tibble: 1 x 1
    pval
  <dbl>
1 0.04
```

This means that our empirical Bootstrap p-value[3] is about 0.04, as opposed to the traditional p-value of 0.16 rooted in statistical assumptions. This is helpful because people are often familiar with statistical p-values, and Bootstrap p-values can be used instead. From a business perspective, we can now be confident that the regression coefficient is between null and strongly negative. In addition, we could easily calculate the equivalent of a p-value for any other threshold (e.g., if we wanted to use −1 instead of zero as a threshold), or for any interval, such as [−1; +1], if we wanted.

## When to Use the Bootstrap

Hopefully, by now you're convinced of the virtues of the Bootstrap for small and oddly shaped data sets. But what about large or evenly shaped data sets? Should you always use the Bootstrap? The short answer is that it is never wrong to use it, but it can be impractical or overkill. For experimental data, we'll rely extensively on the Bootstrap, as we'll see in Part IV of the book. For observational data analysis, which is the focus of this chapter, things are more complicated. Figure 7-7 presents the decision tree we'll use. It may look a bit intimidating, but it can be broken down conceptually into three blocks:

---

2 To see why, note that if you have a 90%-CI, you'll have 5% of values remaining on each side out of it because $(1 − 0.9) / 2 = 0.05$. Conversely, if you see that you have 5% of values on one side out of a CI, then it's the 90%-CI, because $(1 − 2 * 0.05) = 0.9$.

3 To be fully accurate, our Bootstrap p-value would better be called the Bootstrap achieved significance level (ASL).

- If you only want a central estimate (e.g., a regression coefficient) and the conditions for the traditional estimate to be sufficient are fulfilled, you can use it.
- If you want a CI and the conditions for the traditional CI to be sufficient are fulfilled, you can use it.
- In any other case or when in doubt, use the Bootstrap CI.

Let's review these blocks in turn.

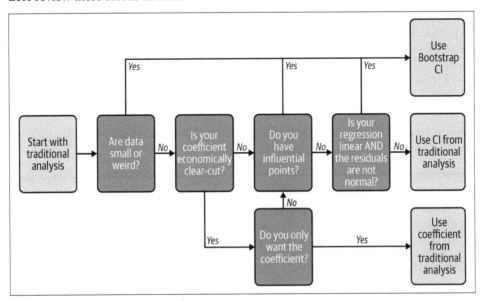

*Figure 7-7. Decision tree to use the Bootstrap*

## Conditions for the Traditional Central Estimate to Be Sufficient

The first thing to keep in mind is that the Bootstrap yields a central estimate or coefficient that is very close to the one obtained by traditional methods (i.e., whatever you would have done if you didn't know about the Bootstrap). Therefore, it never makes sense to start directly with the Bootstrap, when a traditional estimate is one line of code away.

However, if your data is small (typically less than 100 rows) or in any regard weird (e.g., it has multiple peaks or is asymmetric), then that central estimate can be misleading. In that case, you should really use the Bootstrap to calculate a confidence interval instead, ideally displaying its results in the form of a histogram as we did in Figure 7-6.

Similarly, if the coefficient is close to a boundary or threshold, and therefore not economically clear-cut, you'll need to use a CI and the central estimate won't be sufficient.

Even when things are as clean and clear-cut as they can be, you may still want to have a CI, for example because your boss or your business partner required it.

## Conditions for the Traditional CI to Be Sufficient

If you want to have a CI but your data is not so small or weird that the Bootstrap CI is required, the question becomes whether a traditional CI would be reliable and sufficient for your purpose. There are two tests you need to run in that situation:

- Check the presence or not of influential points.
- Check the normality of the regression residuals (only if the regression is linear).

Only if your data is devoid of influential points and you don't see any issue with the residuals can you use the traditional CI.

Influential points are points whose deletion would substantially modify the regression, and there is a statistic, Cook's distance (*https://oreil.ly/0OS4s*), which measures precisely that. For our purposes here, it's enough to know that a data point is considered influential if its Cook's distance is more than one. R and Python have one-liners to calculate Cook's distance for points with respect to a regression model:

```
## Python (output not shown)
CD = st_inf.OLSInfluence(lin_mod).summary_frame()['cooks_d']
CD[CD > 1]

## R
> CD <- cooks.distance(mod)
> CD[CD > 1]
      10
1.45656
```

By definition, an influential point doesn't follow the same pattern as the other points (otherwise deleting it would not drastically change the results of our regression). This means that an influential point is always an outlier, but an outlier is not always an influential point: an outlier lies far out from the cloud that the other points form, but it could still be close to the regression line calculated without it and have a small Cook's distance. In our baking example, the outlier point is also an influential point.

If you have any influential point in your data, it suggests that the standard distributional assumptions are not met, and using the Bootstrap may be wiser.

If you don't have influential points in your data, there is a second check you need to do in the case of a linear regression: you need to make sure that the regression residuals are approximately normal. This doesn't apply to logistic regression because its residuals follow a Bernoulli distribution and not a normal distribution. This check answers both the questions of "How non-normal is non-normal?" and "How large is large?" because they are related: larger data dampens minor deviations from normality, so a degree of non-normality that would be problematic with a hundred points might be OK with a hundred thousand.

Let's extract the regression residuals and visually assess their normality. In R, we obtain the residuals by applying the function resid() to our linear regression model:

```
## R
res_dat <- tibble(res = resid(mod))
p1 <- ggplot(res_dat, aes(res)) + geom_density() + xlab("regression residuals")
p2 <- ggplot(res_dat, aes(sample=res)) + geom_qq() + geom_qq_line() +
    coord_flip()
ggarrange(p1, p2, ncol=2, nrow=1)
```

The syntax in Python is also straightforward: we first get the residuals from the model, then draw a density plot from the Seaborn package, and draw the QQ-plot with the statsmodels package:

```
## Python
res_df = lin_mod.resid
sns.kdeplot(res_df)
fig = sm.qqplot(res_df, line='s')
plt.show()
```

Figure 7-8 displays the two plots we created in R, a density plot and a QQ-plot.

Let's first look at the density plot on the left. For a normal density, we would expect to see a curve with a single peak centered on zero and with smoothly decreasing, symmetric left and right tails. This is clearly not the case here due to the presence of an outlier with a large residual, so we conclude that the residuals are not normally distributed.

The plot on the right is a QQ-plot (QQ stands for Quantile-Quantile), plotted with geom_qq() or qqplot(), which shows the values of our residuals on the x-axis and a theoretical normal distribution on the y-axis. For a normal density, we would expect all the points to be on the line or very close to it, which is again not the case here because of the outlier.

Whenever the residuals of a linear regression are not normally distributed, the Bootstrap will give you better results for CIs and p-values than the traditional approach.

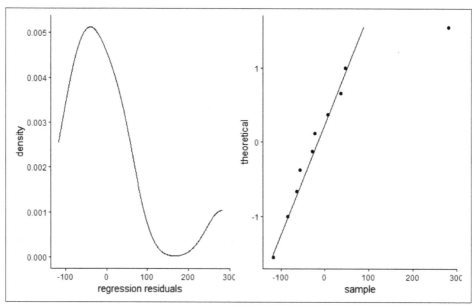

*Figure 7-8. Density plot (left) and QQ-plot (right) of regression residuals*

To recap, it is never wrong to build Bootstrap CIs and you can always fall back on them. But when you only need the central estimate and you can safely rely on it, or when you can safely rely on a traditional CI, it can be overkill to jump to the Bootstrap.

Finally, let's see in a bit more detail how you can determine the number of Bootstrap samples to use.

## Determining the Number of Bootstrap Samples

Once you have decided to use the Bootstrap, you need to determine the number of samples to use in your simulation. If you just want to get a broad sense of the variability of an estimate, B = 25 to 200 gives reasonably robust results for main estimates, according to Efron, the "inventor" of the Bootstrap. Think of it as a 75%-CI. You wouldn't bet the farm on it, but it tells you more than just a mean.

On the other hand, let's say you want a precise p-value or 95%-CI because there is uncertainty as to whether or not a critical threshold, usually zero, is in it or not. Then you'll need a much larger B, because we're typically looking at the 2.5% smallest or 2.5% highest values of the Bootstrap distribution. With B = 200, the lower bound of a two-tailed 95%-CI is equal to 200 * 2.5%, or the fifth-smallest value, and similarly the upper bound is equal to the fifth-largest value. Five is a pretty small number. You can pretty easily get unlucky and get five numbers that are smaller or larger than expected, and throw off your CI bound. Let's visualize that by repeating the Bootstrap regression from the previous section with only 200 samples. As you can see in Figure 7-9, the shape of the distribution is overall similar to Figure 7-5, but now the upper bound for our CI is *above* zero.

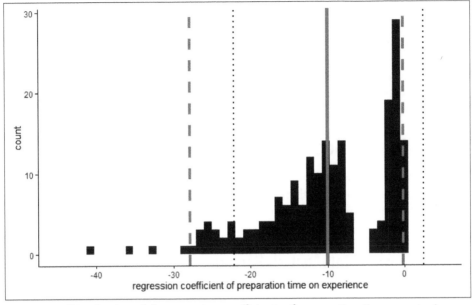

*Figure 7-9. Distribution of the regression coefficients of preparation time on experience, with their mean (thick line), the Bootstrap CI bounds (thick dashed lines) and the normal CI bounds (thin dotted lines) (B = 200 Bootstrap samples)*

Therefore, if a business decision hinges on where that bound is in relation to zero, you'll need to make sure that you estimate it accurately by increasing B. Having 1,000 or even 2,000 samples is a generally accepted guideline in such circumstances. At B = 2,000 the 2.5% quantile is equal to the 50th value, so the odds are much more in your favor. In addition, with very small data sets such as the one used in this chapter, even simulating 4,000 samples takes no more than a few seconds, which is why I have used such a large B.

Let's recap when to use the Bootstrap with observational data, bringing together the test conditions and the number of samples:

- Always start with your traditional regression model to get the main estimate.
- If you have less than 100 points in your data, always use the Bootstrap with B between 25 and 200 to assess the uncertainty around that estimate.
- With N > 100, check your data for signs of influential points (with Cook's distance) or non-normality (with the density plot and QQ-plot of residuals). If

anything looks fishy, use the Bootstrap, again with B between 25 and 200 for main estimates.

- Regardless of N, if you need a precise confidence interval or achieved significance level (a.k.a. p-value), do another Bootstrap simulation with B between 1,000 and 2,000.

- Once you've gotten a sense of how long it takes to run a Bootstrap simulation on your data with a small to medium B, and what the corresponding histogram or CI looks like, always feel free to push the dial on B. Feel free to run a simulation with B = 10,000 overnight to get a nicely dented graph and an exactingly precise CI bound.

# Optimizing the Bootstrap in R and Python

I have shown you how to apply the Bootstrap algorithm "by hand," so that you can understand what it does, but there are packages that will do it in fewer lines of code and will run faster. They will also allow you to use improved versions of the Bootstrap that would be impractical to code manually.

## R: The BehavioralDataAnalysis Package

The companion package to this book includes a boot_ci() function that I have developed with ease of use in mind. In its minimal form, it takes two arguments: a data frame (or tibble) and a function. It will then return the 90% confidence interval of that function on that data frame, calculated on B=100 Bootstrap samples:

```
## R
library(BehavioralDataAnalysis)
my_data <- data.frame(x = rnorm(100), y = rnorm(100))
my_function <- function(df) { return(mean(df$x)) }

CI <- boot_ci(my_data, my_function)
```

Note that the function must take as argument a data frame or tibble and return a single value, which means you might have to wrap a basic R function in your own function as a I did here for the mean.

However, because I advocate so much for the use of regression as the primary method of analysis, I have also added the possibility of passing directly a regression formula instead of a function:

```
CI <- boot_ci(my_data, "y ~ x")
```

Ain't that neat? This makes it possible to calculate confidence intervals on the fly at the end of a dplyr workflow, for instance right after filtering down to some specific date, etc. You can learn more about the function on the GitHub page for the package (*https://github.com/BehavioralDataAnalysis/R_package*).

# R: The boot Package

As a cost of its simplicity, the `boot_ci()` function is pretty basic, and the "naïve" Bootstrap that it calculates is not the most accurate one. If you want more bells and whistles and have the computing power and memory to support them, the boot package and its `boot()` function provide a one-stop shop, "batteries included", for Bootstrap analyses. However, the way it generates Bootstrap samples is not intuitive, so it's worth looking at that feature separately first.

Remember that in the earlier section on Bootstrap for regression analysis, I generated Bootstrap samples with the `slice_sample()` function before running our regression of interest on them:

```R
## R
(...)
for(i in 1:B){
  boot_dat <- slice_sample(dat, n=N, replace = TRUE)
  summ <- summary(lm(times~experience, data=boot_dat))
(...)
```

An alternative approach to generate Bootstrap samples is to take a list of indices, sample from it with replacement, then subset our data based on that list:

```R
## R
> I <- c(1:10)
> I
 [1]  1  2  3  4  5  6  7  8  9 10
> J <- sample(I, 10, replace = TRUE)
> J
 [1] 10  3  1  1  6  1  9  3  4  3
> boot_dat <- dat[J,]
```

This is the approach used in the `boot()` function. We must create a function taking as arguments our original data and a list of index J and returning our variable of interest (here, the regression estimate for experience). The `boot()` function will take care of generating that list for each iteration; we only need to subset our data with it within our function:

```R
## R
boot_fun <- function(dat, J){
  Boot_dat <- dat[J,]
  summ <- summary(lm(times~experience, data=boot_dat))
  coeff <- summ$coefficients['experience','Estimate']
  return(coeff)
}
```

After creating that function, we pass it to the `boot()` function as the argument `statis tic`, as well as our original data as `data`, and the number of Bootstrap samples as R (for replications). The `boot()` function returns an object that we then pass to the `boot.ci()` function to get our CI:

```
## R
> boot.out <- boot(data = dat, statistic = boot_fun, R = 2000)
> boot.ci(boot.out, conf = 0.95, type = c('norm', 'perc', 'bca'))
BOOTSTRAP CONFIDENCE INTERVAL CALCULATIONS
Based on 2000 bootstrap replicates

CALL :
boot.ci(boot.out = boot.out, conf = 0.95, type = c("norm", "perc",
    "bca"))

Intervals :
Level      Normal              Percentile          BCa
95%   (-25.740,   6.567 )  (-28.784,  -0.168 )  (-38.144,  -0.383 )
Calculations and Intervals on Original Scale
```

The boot.ci() function can return a variety of CIs, as determined by the parameter type. "norm" is the traditional CI based on a normal distribution. "perc" is the percentile, or quantile, Bootstrap that we calculated by hand previously. "bca" is the bias-corrected and accelerated percentile Bootstrap ($BC_a$). The $BC_a$ Bootstrap refines the percentile Bootstrap by leveraging some of its statistical properties; these are beyond our scope here. You can learn more about them in any of the sources listed as references; suffice it to say that the $BC_a$ Bootstrap is considered best practice when using Bootstrap simulations. It can be pretty demanding in terms of computations however, so I would recommend using the percentile Bootstrap first, and once you have a reasonably final version of your code, try switching to the $BC_a$ Bootstrap.

In the present case, the normal and percentile CIs are very close to what we calculated by hand, as expected. The $BC_a$ CI shifts to the left, strengthening our original conclusions that the coefficient is most likely strongly negative.

Finally, please note the difference in the naming of the functions in the two packages: my package has a boot_ci() function, with an underscore like functions in the Tidyverse, whereas the boot package chains together its boot() and boot.ci() functions, the latter with a more traditional dot notation.

---

## High-Performance Bootstrap

Using the boot package can typically make your code two to five times faster than with basic loops, but sometimes that's just not enough. Should you need more computational oomph, the Rfast package offers a bare-bones implementation of regression that will give you an additional order of magnitude improvement (e.g., twenty to fifty times faster than our original code). I use it under the hood in my package.

---

# Python Optimization

Python offers very different trade-offs to the analyst compared to R: on the one hand, it has fewer statistical packages and there is no equivalent of the R boot package that would implement the Bootstrap algorithm directly. On the other hand, I find it to be more forgiving of beginners in terms of performance. This is especially true for boot-strapping, because for loops, which beginners tend to use a lot, are comparatively much less costly. Therefore, I expect Python users to get much more mileage out of the naive implementation we started with.

Still, should you need to add computational oomph to your Bootstrap implementation in Python, you can do so by going "full NumPy":

```
## Python
# Creating unique numpy array for sampling
data_ar = data_df.to_numpy()  ❶
rng = np.random.default_rng()  ❷

np_lst = []
for i in range(B):
    # Extracting the relevant columns from array
    boot_ar = rng.choice(data_ar, size=N, replace=True)  ❸
    X = boot_ar[:,1]  ❹
    X = np.c_[X, np.ones(N)]
    Y = boot_ar[:,0]  ❺

    ### LSTQ implementation
    np_lst.append(np.linalg.lstsq(X, Y, rcond=-1)[0][0])  ❻
```

❶ We convert our original pandas dataframe to a NumPy array.

❷ We initialize the NumPy random number generator only once, outside of the loop.

❸ We create our bootstrapped data set by using the NumPy random number generator, which is significantly faster than the pandas .sample() method for dataframes.

❹ We extract the predictor columns from our array and in the following line manually add a constant column for the intercept (whereas statsmodel was previously handling that under the hood for us).

❺ We extract the column for the dependent variable from our array.

❻ Reading the function calls from the right to the left: we fit the linear regression model with the np.linalg.lstsq() function to the predictor and dependent variable data. The rcond=-1 parameter removes an unimportant warning. The value we want is in the [0][0] cell for this particular model; you can find the specific cell you need by running `np.linalg.lstsq(X, Y, rcond=-1)` once and inspecting its output. Finally, we append the value to our result list.

Going full NumPy can significantly improve your performance, to the order of fifty times faster or so for larger data sets. However, our original code did just fine for our small data set, and it was more readable and less error-prone. Moreover, if you go beyond straightforward linear or logistic regressions, you'll have to search on the Internet for a NumPy implementation of the algorithm you want. But should you need to improve the performance of your Bootstrap code in Python, you now know how to do so.

## Conclusion

Behavioral data analyses often have to deal with smaller or weirder data. Fortunately, the advent of computer simulations has given us in the Bootstrap a great tool to deal with such situations. Bootstrap confidence intervals allow us to correctly assess the uncertainty in our estimates without relying on potentially faulty statistical assumptions about the distribution of our data. With observational data, the Bootstrap is most useful when our data exhibits signs of influential points or non-normality; otherwise, it is often overkill. With experimental data, however, the heavy reliance on p-values to make decisions means that we'll use it extensively, as we'll see in the next part of the book.

# PART IV

# Designing and Analyzing Experiments

Running experiments is the bread and butter of behavioral scientists and causal data scientists in business. Indeed, randomizing the allocation of subjects between experimental groups allows us to negate any potential confounding without the need to even identify it.

Books about A/B testing abound. How is the presentation in this one different? I would argue that several aspects of its approach make it both simpler and more powerful.

First of all, recasting experiments within the causal-behavioral framework will allow you to create better and more effective experiments, and better understand the spectrum from observational to experimental data analysis instead of thinking of them as separate.

Second, most books on A/B testing rely on statistical tests such as the T-test of means or the test of proportions. Instead, I'll rely on our known workhorses, linear and logistic regressions, which will make our experiments simpler and more powerful.

Finally, traditional approaches to experimentation decide whether to implement the tested intervention based on its p-value, which doesn't lead to the best business decisions. Instead I'll rely on the Bootstrap and its confidence intervals, which is progressively establishing itself as the best practice.

Therefore, Chapter 8 will show what a "simple" A/B test looks like when using regression and the Bootstrap. That is, for each customer we toss a metaphorical coin. Heads and they see version A, tails and they see version B. That is often the only solution for website A/B testing.

However, if you know ahead of time the people from which you'll draw your experimental subjects, you can create more balanced experimental groups with stratification. This can significantly increase the power of your experiments, as I'll show in Chapter 9.

Finally, it happens quite often that you can't randomize at the desired level. For example, you're interested in the impact of a change on customers, but you must randomize at the level of call-center representatives. This requires cluster randomization and hierarchical modeling, which we'll see in Chapter 10.

# CHAPTER 8

# Experimental Design: The Basics

Let's start our exploration of experimentation with a very simple experiment: influenced by a leading online store, AirCnC's management has decided that a "1-click booking" button is just what's needed to boost AirCnC's booking rate. As I discussed earlier, we'll assign customers to our experimental groups one by one as they connect to the website. This is the simplest possible type of experiment, and many companies offer interfaces that allow you to create and start running A/B tests like this in a matter of minutes.

This straightforward experiment will be the opportunity to go through the process without getting bogged down in technical considerations:

1. The first step is to plan the experiment. This is where the causal-behavioral perspective comes in, to help ensure that you have clearly defined criteria for success, and that you understand what it is you're testing and how you expect it to impact your target metric.

2. Then, after reviewing the data and packages that we'll use in the rest of the chapter, I'll show you how to do the random assignment and determine the sample size for your experiment.

3. Finally, we'll analyze the results of the experiment, which in such a simple case will be very quick.

The vocabulary of experimental design owes much to its statistical and scientific roots. I'll talk of "control" and "treatment" groups as well as "interventions," which may sound ominous or like overkill when we're really discussing the position of a button on a website or the amount of a discount. When talking about experiments in general, there isn't much I can do to use a simpler vocabulary; but when you're talking to your business partners about a specific experiment, I would encourage you to stick with concrete terms relating to that experiment (e.g. "the old and the new creatives," "the group with the lower discount and the group with the higher discount," etc.).

Anecdote: once when I suggested an "intervention," a business partner thought I meant that they weren't doing their job well and I needed to intervene. Not the best start for a fruitful and trusting relationship. Meet people where they are and make an effort to speak their language instead of expecting them to know yours.

## Planning the Experiment: Theory of Change

Planning is a crucial step of experimental design. Experiments can fail for a variety of reasons, many of which you can't control, such as the implementation going awry; but poor planning is both a frequent cause of failure and one you *can* control. Anyone who runs experiments for a living has horror stories of experiments that may or may not have been technically impeccable but were utterly pointless because people didn't have clarity about what was tested.

At the end of the day, no process can save you if you're just going through the motions without exercising your business acumen and common sense, but hopefully the formula I'm going to outline will help you make sure you have covered all your bases. We'll borrow a concept from nonprofit and governmental planning, namely the *theory of change* (ToC). In one sentence, your ToC should connect what you're doing to your ultimate business goal and target metric through a behavioral change:

> Implementing [INTERVENTION] will help us achieve [BUSINESS GOAL], as measured by [TARGET METRIC], through [BEHAVIORAL LOGIC].

I'll elaborate on each of the four components in turn, but to give you a sense of where we'll land, here's what our final theory of change will look like:

> Implementing [a 1-click booking button] will help us achieve [higher revenue], as measured by [booking probability], through [a reduction in the duration of the booking process].

This can be represented in CD format, as in Figure 8-1.

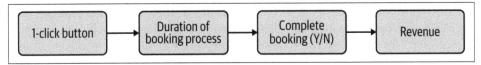

*Figure 8-1. Theory of change for our experiment*

Let's first review our business goal and target metric, then our intervention, and finally our behavioral logic.

## Business Goal and Target Metric

You might be surprised that I start with the business goal and target metric instead of the definition of the intervention. After all, shouldn't we know what we're testing first? Unfortunately, a common cause of failure is the decision to test something (often the latest management fad or something that your boss's boss read about) without a clear sense of what we're trying to achieve.

### Business goal

The first step is to determine the business goal for the experiment. Companies are usually trying to increase their profit, but just putting "higher profit" as your business goal would not be very helpful. Instead, I would recommend going one level deeper and using a variable such as revenue, costs, customer retention, etc., that is more concrete but of obvious benefit to the company. This may seem a trivial step, but it can actually surface disagreements about the goal of an experiment (e.g., is it to reduce costs or increase revenue?). Here, the business goal for the 1-click button experiment is higher revenue (Figure 8-2).

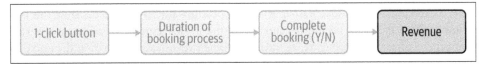

*Figure 8-2. Our business goal is revenue*

### Target metric

The second step is to decide how you'll measure success at the end of the experiment, i.e., your target metric. There is a trade-off at play here: on the one hand, you want to use a metric as close to profit as possible, such as dollars of additional revenue or decreased cost; on the other hand, you want to pick a metric as close to your intervention as possible, to reduce extraneous noise.

A compromise you'll often have to make at this point is to use "leading indicators"— basically causes of the variables you're ultimately interested in. For example, you may ultimately care about your customers' LTV (lifetime value, the total amount they'll spend with your company) but have to settle for a three-month booking amount. Similarly, sign-up can be used as a leading indicator for usage, usage for amount purchased, etc. This will allow you to report results much earlier than if you used long-term business metrics, while still having a clear connection to your business goal.

However, if your target metric is an operational metric instead of a financial metric, things can get hairy. If the button shortens how long it takes someone to book a trip but otherwise doesn't increase bookings in any way, is that a success or not? What about satisfaction with the booking experience and net promoter score? These may not translate directly into dollar figures, but at the same time it is not unreasonable to assume that improving them has a positive impact for your business. This is sometimes done informally, by picking operational metrics as targets for experimentation and assuming with some hand-waving that these will end up benefiting the company's bottom line. Of course, armed with this book, we can do better. We can validate and measure the causal connection between a short-term operational metric and long-term business outcomes through an observational study or a dedicated experiment, as we'll see later.

### Pitfalls of poor target metrics

The goal here again is to make good business decisions. I don't want to be a fanatic of dollar figures, because it would be unduly restrictive and would exclude a large range of business improvements. At the same time, you want to make sure that you have a measurable target metric that you'll be able to track. There are several potential pitfalls that you should avoid here.

The first one is picking something you can't measure reliably. What would it mean to say, "The 1-click button makes the booking experience easier"? How would you measure it? Asking the product manager or the product owner to decide after the fact whether there has been an improvement is not a measurement. "Our customers rate the website as easier to navigate, as measured by a two-question survey at the end of a visit" may be an imperfect proxy but at least it can be measured. This is why it makes sense to express the business goal and the target metric separately: the former expresses your true intent, even if it's not measurable, while the latter shows clearly what you intend to measure. This avoids misunderstandings and moving goal posts.

The second pitfall is to write down a laundry list of metrics, such as "success would be an improvement in booking rate, booking amount, customer satisfaction, or net promoter score," or even worse, to wait until after seeing the results to determine the metrics for success—e.g., you were originally thinking that the experiment would improve customer experience, but when the results come, customer experience is flat and average website session duration has improved. The problem with that approach is that it increases the risk of false positives (calling the result a success when it was a random fluke).[1] It's okay however to have up to two or three target metrics that are clearly defined before the experiment as long as you take that into account when analyzing the results (more on that later). Some people advocate using a composite of multiple metrics (e.g., a weighted average), which is called an *Overall Evaluation Criterion* (OEC), but I personally feel that it often obscures things more than it helps. I would rather encourage you to clearly articulate your theory of change and how the various metrics are related to each other—for example, do you expect the 1-click button to improve booking rate *and* customer experience, or booking rate *through* customer experience?

To conclude, in the case of the 1-click button experiment, we could directly use booking revenue as our target metric, but we don't expect that our intervention will modify the average amount per booking, so it makes more sense to use the probability that a customer will complete a booking (Figure 8-3).

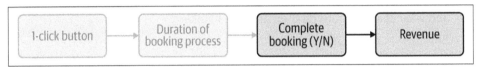

*Figure 8-3. Adding the target metric, the probability of completing a booking*

## Intervention

Once we have our business goal and target metric, we can work on defining our intervention. Here the idea for the "1-click button" intervention comes from the company's management (Figure 8-4), but it could also have come from UX or behavioral research: identifying issues and opportunities for improvement in a company's processes, products, and services is indeed one of the main tasks of researchers in business but it's outside the scope of this book.

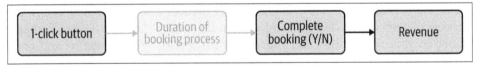

*Figure 8-4. Adding the intervention, the 1-click button*

On the face of it, what could be simpler than a "1-click booking" button? Most of us have seen it implemented on the website of an online retailer or another and the idea seems perfectly straightforward. But there can be a large gap between a business idea,

---

1 This is why pharmaceutical trials, as well as an increasing number of social science experiments, are preregistered. You can't set out to test a drug for a heart condition and then decide after the fact that it is an effective drug against hair loss because the patients in the treatment group have seen their hairline move forward; you'd need to run a second experiment, itself preregistered, for the purpose of testing the new hypothesized effect.

so used and familiar that everyone immediately feels they know what it is, and a specific implementation. If you think about the nitty-gritty details of how it would be implemented, there are actually a lot of questions to answer, each with multiple possible answers:

- At which point in the process does the button become available?
- Where is the button located?
- What does the button look like? Is it in the same colors as the other buttons on the page or does it stand out in a bright and rich color?
- What is written on the button?
- What information do we need about a customer to make 1-click booking available and how do we make sure we have it?
- What happens after the customer clicks the button? What page are they taken to, and what, if any, actions do they still have to take?
- Etc.

Let's say for example that the website's color theme is pastel green and blue, to hint at nature and travel, and the new button is a shiny red. If that button increases booking, it might be because of the attractiveness of 1-click booking, but it could also be because customers otherwise struggle to see the normal booking button and give up. In that case, the cause of the increase in booking is really "a more visible booking button" rather than "a 1-click booking button," but you can't distinguish between the two because the 1-click button was also more visible. Unfortunately, you can't ever just test one idea, as you're always also testing the many aspects of how it is implemented.

The lesson here is that A/B testing is a powerful but narrow tool. You need to be careful not to make—or let others make—grand statements about what a specific experiment says. It is definitely easier said than done, because business partners often want answers that are broad, clear-cut, and without fine print. With that said, even simply stating in your presentation that you're testing a specific implementation and not a general idea can be useful, for example, "this experiment will test the impact of a 1-click button under such and such conditions, and its results should not be interpreted as applying to booking buttons more broadly."

More generally, I would recommend you test the smallest possible intervention you can. In this case, you could try changing the color or position of the booking button before implementing the bigger change that is a 1-click button. You might get pushback from your business partners, who often want to run an "omnibus" test with multiple changes at once; in that case, make clear that you're really just testing that none of the changes break the experience, as opposed to actually measuring impacts. A better alternative would be to test different implementations of the same concept in the experiment. If four slightly different 1-click button treatments have the same impact, you can more confidently draw conclusions regarding the general impact of 1-click booking; on the other hand, if they have very different impacts, then this suggests that implementation matters a lot and that you need to be very careful with your conclusions regarding how a specific implementation will generalize.

## Behavioral Logic

Once we know our business goal and target metrics and we have defined our intervention, the last step is to connect the two through the behavioral logic of your theory of change: why and how would our intervention impact our target metric?

This is another surprisingly common source of failure for experiments: a problem has been identified and someone decides to implement the latest management fad that they have been thinking about recently, even though it's unclear why it would help with this specific problem. Or someone decides that a more attractive and simpler user interface (UI) will increase purchase amounts. To have confidence in our experiment, you need to be able to articulate a reasonable behavioral story.[2] In the case of 1-click booking, you might hypothesize that customers identify an attractive booking but give up before completing their booking because the booking process is cumbersome; the 1-click button would impact the probability of booking by shortening and simplifying the booking process. This is typically where your ToC comes together in a CD, in this case the one I showed you at the beginning of the chapter (Figure 8-5).

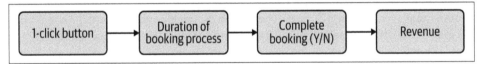

*Figure 8-5. The complete CD for our theory of change*

Overall, articulating your behavioral logic has two benefits. First, it is often testable in itself. Do a large number of customers actually drop off between starting a booking and completing it? If so, the hypothesized logic makes sense from a behavioral data perspective. But if, for example, most customers leave without having started the booking process with a specific product, e.g., because they couldn't find something that appealed to them or they were overwhelmed with the number of options, then AirCnC is trying to solve the wrong problem; it's unlikely that offering 1-click booking will improve their numbers.

Ask yourself: what would confirm or refute our logic? What would the data look like if it were true or false? If you don't have the necessary data at hand, it might be worth running a preliminary test, such as bringing in 10 people for user experience testing,

---

2 Wendel (2020) is a great resource to understand the obstacles and drivers of a behavior and build a strong behavioral logic.

e.g., observing them trying to use your website while they're thinking out loud. This may not fully confirm or refute your logic, but it will probably give you some indication, at a fraction of the development cost for the solution considered. As the often-cited quote by Albert Einstein goes, "If I had an hour to solve a problem, I'd spend 55 minutes thinking about the problem and 5 minutes thinking about solutions."

The second benefit of articulating the behavioral logic of your intervention is that it will generally give you a sense of the potential upside. What would the numbers look like, in the best-case scenario, if the problem at hand was solved? Assuming that all customers who drop off during the booking process would complete it with 1-click booking, how much would the booking rate increase? This is the best-case scenario from the perspective of the experiment because we're assuming that it would fully solve the problem, which is unlikely in reality. If the increase in booking rate under this scenario would not pay for the implementation of the 1-click booking, then don't even bother testing it.

Once you've validated that your best-case scenario would be profitable, you can start thinking about your most likely scenario. How much would we expect 1-click booking to improve the booking rate? There is undoubtedly a lot of subjectivity and uncertainty involved, but having articulated the behavioral mechanism at play, you can generally make reasonable guesses. Do you really expect that 75% of people are dropping off the booking process because it's taking too long? In addition, it can be a worthwhile exercise by making explicit people's assumptions and gut feelings. If the product manager and the UX researcher vastly disagree on the percentage of customers who drop off because the process is taking too long, you need to close that gap first. What does one of them know that the other doesn't? Use your business sense and understanding of the processes. If most customers are dropping off at the exact same step in the process, e.g., payment, then it's likely that there is something wrong with that specific step—people don't all run out of patience at the exact same time. You can then compare the expected benefits with the cost of implementing the solution. Is it still worthwhile?

You can also approach this question from the other side: start by determining the breakeven point of the solution, i.e., the improvement in the target metric that would make the solution profitable to implement, and then consider whether that improvement is realistic from a behavioral standpoint. From a psychological perspective, it's better to start with the expected result than with the breakeven point: if you start with the breakeven point, you're more likely to anchor on it and find reasons to justify that it's achievable. However, in many cases, you'll know the breakeven point first, for example if it was calculated during a preliminary cost-benefit analysis; your company or business partners might also request it and refuse to think about the expected result first. Don't worry too much about it. Regardless of whether you're working with your expected or best-case result, we'll need it to determine the minimum detectable effect for our experiment.

It's important that your behavioral logic connects your proposed solution with your target metric. Don't leave it to "this will improve the customer experience." How will you know it? If your logic is solid, you should be able to express it in terms of a causal diagram, with at least some of the effects being observable.

A useful rule of thumb to help you articulate your logic is to break down your business metric into components. For example, revenue (or most of its variations) can be broken down into number of customers, probability/frequency of purchase, quantity purchased, and price paid. Determining which components are likely to be affected can allow you to better articulate the business case. If your business partners are concerned by a decrease in the number of customers and the proposed intervention would most likely only increase the quantity purchased, you need to clarify with them that they would still call it a win. This approach can also reduce the noise in your experiment; if the proposed intervention would most likely only increase the quantity purchased, you can focus on that metric and disregard somewhat random fluctuations in price paid.

# Data and Packages

The GitHub folder for this chapter (*https://oreil.ly/BehavioralDataAnalysisCh8*) contains two CSV files with the variables listed in Table 8-1. In the table, the check mark (✓) indicates the variables present in that file, while the cross (✗) indicates the variables missing.

*Table 8-1. Variables in our data*

|  | Variable description | chap8-historical_data.csv | chap8-experimental_data.csv |
|---|---|---|---|
| *Gender* | Categorical, "male"/ "female" | ✓ | ✓ |
| *Period* | Month index, 1-32 in historical data, 33 in experimental data | ✓ | ✓ |
| *Seasonality* | Annual seasonality, between 0 and 1 | ✓ | ✓ |
| *Month* | Month of year, 1-12 | ✓ | ✓ |
| *Booked* | Binary 0/1, target variable | ✓ | ✓ |
| *Oneclick* | Binary 0/1, experimental treatment | ✗ | ✓ |

In this chapter, we'll use the following packages in addition to the standard ones called out in the Preface:

```
## R
library(pwr) # For traditional power analysis

## Python
import statsmodels.stats.proportion as ssprop # For the standardized effect size
import statsmodels.stats.power as ssp # For traditional power analysis
```

# Determining Random Assignment and Sample Size/Power

Once you've built and validated the theory of change of your experiment, the next step is to determine how you'll do the random assignment and how big a sample size you'll need.

In my experience, this is often a big step the first time you're running an experiment in a certain environment. Taking a serious look at your historical data often yields surprising insights that can reshape an experiment. In addition, depending on the noise in your data and the expected impact size (small if you've done your job correctly of defining a narrow scope for your experiment), discovering how large a sample you'll need can be humbling. I still remember the first time the numbers came back and I was told that we would need to run an experiment for almost a year.

## Random Assignment

The theory of random assignment could not be simpler: whenever a customer reaches the relevant page, they should be shown the current version of the page (called "control" in experimental jargon) with a certain probability and the version with the 1-click booking button (the "treatment") with the opposite probability.

The most straightforward option is a 50%-50% allocation until you've reached your target sample size, but you may want to use a different split if you have a very high volume of transactions. Let's imagine for example that you're managing a website with 100 million visits a day, and you have determined that your necessary sample size is 2 million. You could simply go with 50%-50%, and be done with your experiment in about 30 minutes. However, if anything goes wrong with your treatment (e.g., a bug crashes the website, admittedly an extreme case), you'll have 1 million unhappy customers on your hands before you know it. In addition, maybe customers during these 30 minutes are not representative of your full customer base (e.g., China is asleep at that time and you get mostly American visitors or vice versa). In such a situation, it would be better to get the 1 million visits you want in your treatment group over the course of a more representative period, such as a week or a month (you don't have to worry about the control group being bigger than 1 million). For a 100 million visits/day website, that would translate into splits of respectively 99.86%-0.14% (because 1 / (7 * 100) = 0.14%) and 99.97%-0.03% (because 1 / (30 * 100) = 0.03%). For the sake of simplicity, I'll assume a 50%-50% split in the rest of the chapter.

## Code implementation

From a coding perspective, assuming that you are not using a software that takes care of it for you, this can easily be implemented in R or Python:

1. Whenever a new customer reaches the relevant page, we generate a random number between 0 and 1.

2. Then we assign the customer a group based on that random number: if K is the number of groups we want (including a control group), then all individuals with a random number less than 1/K are assigned to the first group; all individuals with a random number between 1/K and 2/K are assigned to the second group, and so on.

Here, K is equal to 2, which translates into a very simple formula:

```
## R
> K <- 2
> assgnt = runif(1,0,1)
> group = ifelse(assgnt <= 1/K, "control", "treatment")

## Python
K = 2
assgnt = np.random.uniform(0,1,1)
group = "control" if assgnt <= 1/K else "treatment"
```

## Pitfalls of random assignment

There are, however, a certain number of subtleties that can trip novice experimenters. I'll cover two in this chapter: the timing and the level of the assignment.

**Random assignment timing.**   The first one is determining the right point in the process for random assignment. Let's say that whenever a customer gets on the first page of the website, you assign them to either the control or the treatment group. Many of these customers will never reach the point of making a booking and will therefore not see your booking interface. This would dramatically reduce the effectiveness of your experiment because you would in effect experiment only on a fraction of your sample.

When determining which customers should be part of the experiment and when they should be assigned to an experimental group, you should reflect on how the treatment will be implemented if the experiment is a success. Your experimental design should include the same people who would see the treatment if it got implemented in business as usual and only them. For instance, visitors who leave the website before booking will still not see the button, whereas any future promotion or change to the booking page would be built in addition so to speak "on top of" the button, which would always be present. Therefore, nonbooking visitors should be excluded but customers with a promotion should be included.

**Random assignment level.** The second challenge is making sure that the random assignment is happening at the "right" behavioral level. I'll explain what this means with an example. Let's say that a visitor comes on the AirCnC website and starts a booking but then for whatever reason leaves the website (they got disconnected, it's time for dinner, etc.) and comes back to it later. Should they see the same booking page? If they were offered 1-click booking the first time, should they still be offered it the second time?

The problem here is that there are really multiple levels that could potentially make sense. You could assign control or treatment at the level of a single website visit, of a booking, however many visits it takes, or at the level of a customer account (which may or may not be the same person if several people in a household use the same account). Unfortunately, there are no hard rules here, the right approach must be determined on a case-by-case basis by thinking about the conclusions you want to draw and what a permanent implementation would look like.

In many cases, it makes sense to do an assignment at the closest level you can to a human being: customer account if you can't distinguish people in a household, or individual customer if they each have a subaccount, as is the case for example with Netflix. Human beings have persistent memories and alternating options for the same person can get confusing. Here, this would mean that our AirCnC customer should see the 1-click button for the whole duration of the experiment, regardless of how many visits and bookings they make during that time. Unfortunately, this means that you can't just roll the dice metaphorically each time someone starts a booking on the website to determine their assignment; you need to keep track of whether they have been assigned in the past and if so to which group. For a website experiment, this can be done through cookies (assuming the customer allows them!).

 The level at which you're making your random assignment should also be the level at which you calculate your sample size. If you make assignments at the customer level and customers make an average of three visits per month, you'll need a three times longer experiment than if you make assignments at the visit level. But the level you choose for your random assignment should determine your sample size, not the other way around!

Whatever level you choose, you'll have to keep track of the assignment(s) to be able to link them to business outcomes later. This is why the best-in-class approach is to use centralized systems that record all assignments and connect them to customer IDs in a database, so that they can serve a consistent experience to customers over time.

More broadly, what these two challenges point at is that the implementation of a business experiment is almost always a complex technical affair. A variety of vendors now offer somewhat plug-and-play solutions that hide the complexity under the hood, especially for website experimentation. Whether you rely on them or on your internal tech people, you'll need to understand how they're doing the random assignment to make sure that you're getting the experiment you want.

A good way to check that the system works correctly is to start with an A/A test, where there is a random assignment but the two groups see the same version of the page. This will allow you to check that there are indeed the same number of people in the two groups and that they don't differ in any significant way.

## Sample Size and Power Analysis

Once we know what we're going to test and how, we need to determine our sample size. In some cases, as here with our 1-click booking experiment, we can choose our sample size: we can just decide how long we want to run the experiment. In other cases, our sample size might be defined for us, or at least its maximum. If we're going to run a test across our whole customer or employee population, we can't increase that population just for the sake of experimentation!

Regardless of the situation we're in, we will look at our sample size in relation to other experimental variables, such as statistical significance, and not in a vacuum. Understanding how these variables are related is crucial to ensure that our experimental results are usable and that we're drawing the right conclusions from them. Unfortunately, these are highly complex and nuanced statistical concepts and they have not been developed for business decisions.

In line with the spirit of this book, I'll do my best to explain these statistical concepts and conventions in the context of business decisions. I'll then share my reservations about the traditional conventions and offer my two cents regarding how to tweak them while remaining in the traditional framework. And finally, I'll describe an alternative approach that has been slowly gaining momentum and which I think is superior, namely using computer simulations.

### A little bit of statistics theory without math

When running an experiment such as the "1-click booking" button, our goal is to make the right decision: should we implement it or not? Unfortunately, even after running an experiment (or a hundred), we can never be 100% sure that we're making the right decision, because we have only partial information. Certainly, if we ran an experiment for years on end, we might reach a point where there is only one chance in a million that we're wrong, but never exactly zero. Moreover, we generally don't want to run an experiment for years on end, when we could be running other experiments instead! Therefore, there's a trade-off between the sample size of an experiment and our degree of certainty.

Because we can never know after the fact whether a specific decision was right or not, our approach will be to try to select a good sample size and a good decision rule before the experiment. What does "good" mean here? Well, the very best possible sample size and rule would be those that maximize our expected profit over time. The corresponding calculations are doable but require advanced methods that are beyond the scope of this book.[3] Instead, we'll rely on the following measures:

- Assuming that our 1-click button does increase our booking rate, what is the probability that we'll rightly implement the button? This is called the "true positive" probability. On the other hand, if there is a positive effect and we wrongly conclude that there is none, it's called a "false negative."

- Assuming that our 1-click button has no discernible effect (or, God forbid, a negative effect!) on our booking rate, what is the probability that we'll wrongly implement the button? This probability is called the "false positive" probability. On the other hand, if there is no effect and we rightly conclude that there is no effect, it's called a "true negative."

These various configurations are summarized in Table 8-2.

*Table 8-2. Making the right decision in the right situation*

| | | Do we implement the 1-click booking button? | |
|---|---|---|---|
| | | YES | NO |
| Does the 1-click booking button increase the booking rate? | YES | True positive | False negative |
| | NO | False positive | True negative |

We would want our true positive and our true negative rates to be as high as possible, and our false positive and false negative rates to be as low as possible. However, the simplicity of this table is deceptive, and it actually encompasses an infinite number of situations: when we say that the button increases the booking rate, it could mean that the increase is 1%, 2%, etc. On the other hand, when we say that the button does not increase the booking rate, it could mean that it has exactly zero effect, or that it is decreasing the booking rate by 1%, 2%, etc. All these effect sizes would have to be factored in to calculate the overall true positive and true negative rates, which would be too complicated. Instead, we will rely on two threshold values.

The first one is an impact of exactly zero for all subjects, also called the "sharp null hypothesis" (the nonsharp null hypothesis would be an *average* zero effect across subjects). The false positive rate for this value is called the statistical significance of our

---

3 In case you're wondering, you would need to use Bayesian methods. Maybe I'll get to that in the next edition of this book! In the meantime, *Think Bayes* (O'Reilly) by Allen Downey is one of the most approachable introductions I know to the topic.

experiment. Because a negative impact would be easier to catch than a null effect, the false positive rate for any negative value will be at least as large as the statistical significance, and larger negative effects will have higher false positive rates. The most common convention in academic research is to set the statistical significance at 5%, although in certain fields such as particle physics (*https://oreil.ly/U48vk*), it can sometimes be as low as 0.00005%.

The second threshold value is set at some positive effect that we're interested in measuring. For example, we might say that we want to choose a sample size so that we can be "reasonably sure" that we will capture a 1% increase in booking rate, but we're OK with missing smaller effects than that. This value is often called the "alternative hypothesis," and the true positive rate for this value is called the statistical power of our experiment. Because larger effects would be easier to catch, the true positive rate for any larger value will be at least as large as the statistical power, and larger positive effects will have higher true positive rates. "Reasonably sure" is traditionally taken to mean 80%. To reiterate, this doesn't mean that your experiment "has a power of 80%" and that phrase is actually meaningless by itself: the experiment also has a power of 90% for a certain larger effect size, and a power of 70% for a certain smaller effect size, and so on.

Our table updated according to the traditional convention would therefore look like Table 8-3.

*Table 8-3. The thresholds used in the traditional statistical approach*

| | Do we implement the 1-click booking button? | |
| --- | --- | --- |
| | YES | NO |
| The 1-click booking button increases the booking rate by more than 1%. | > 80% (larger for larger effect sizes) | < 20% (smaller for larger effect sizes) |
| **The 1-click booking button increases the booking rate by exactly 1%.** | 80% (statistical power) | 20% (1 minus statistical power) |
| The 1-click booking button increases the booking rate by less than 1%. | < 80% (larger for larger effect sizes) | > 20% (smaller for larger effect sizes) |
| **The 1-click booking button has exactly no impact on the booking rate.** | 5% (statistical significance) | 95% (1 minus statistical significance) |
| The 1-click booking button strictly decreases the booking rate. | < 5% (smaller for larger negative effect sizes) | > 95% (larger for larger negative effect sizes) |

I am no big fan of using an arbitrary number purely because it's conventional, and you should feel free to adjust the "80% power" convention to fit your needs. Using a power of 80% for your relevant threshold effect size would mean that if the intervention had exactly that effect size, on average, you would have a 20% chance of not implementing the intervention because you wrongly got a negative result. For big and costly interventions that are hard to test, my opinion is that it's too low and I would personally target 90% power. On the other hand, the higher the power you want, the larger your sample size will need to be. You may not want to spend half a year getting absolutely certain of the value of the 1-click button if in that time your competitor has completely revamped their website twice and is eating your lunch.

In my personal experience, one key but often ignored consideration for power analysis and sample size determination in the real world is organizational testing velocity: how many experiments can you run in a year? In many companies, that number is constrained by someone's time (either the analyst's or the business partner's), by the company's planning cycle, by budget limits, etc., but not by the number of customers available. If you can realistically hope to plan, test, and implement only one intervention per year, do you really want to run a three-month experiment and then do nothing for the rest of the year? On the other hand, if you can run one experiment a week, do you really want to spend three months getting certain of a positive but mediocre impact instead of taking 12 chances at a big one? Therefore, after doing the math, you should always do a sanity check of your experiment duration based on your testing velocity and adjust it appropriately.

Regarding statistical significance, the conventional approach introduces an asymmetry between the control and the treatment with a statistical significance threshold of 95%. The bar of evidence the treatment has to pass to get implemented is much higher than for the control, which is implemented by default. Let's say that you're setting up a new marketing email campaign and you have two options to test. Why should one version be given the benefit of the doubt over the other? On the other hand, if you have a campaign that has been running for years and for which you have run hundreds of tests, the current version is probably extremely good and a 5% chance of wrongly abandoning it might be too high; the right threshold here might be 99% instead of 95%. More broadly, relying on a conventional value that is the same for all experiments feels to me like a missed opportunity to reflect on the respective costs of false positives and false negatives. In the case of the 1-click button, which is easily reversible and has minimal costs of implementation, I would target a statistical significance threshold of 90% as well.

To recap, from a statistical perspective, our experiment can be summarized by four values:

- The statistical significance, often represented by the Greek letter beta ($\beta$)
- The effect size chosen for the alternative hypothesis, a.k.a. the minimal detectable effect
- The statistical power, often represented as $1 - \alpha$ where $\alpha$ is the false negative rate for the chosen alternative effect size
- The sample size of our experiment, represented by N

---

These four variables are referred to as the B.E.A.N. (beta, effect size, alpha, sample size N), and determining them for an experiment is called a "power analysis."[4] For our 1-click button experiment, we have decided on the first three of them and we only have to determine the sample size. We'll see next how to do it with traditional statistical formulas and then how to do it with computer simulations.

## Traditional power analysis

Statisticians have developed formulas to determine the required sample size for certain statistical tests. Given that we'll rely on regression instead of tests, you might wonder why we would want to use these formulas. In my experience, these will give you values that are of the same order of magnitude as the "true" required sample size. That's a quick and easy way to get reasonable starting values for your simulations if you have no idea whether your sample size should be 100 or 100,000 (in this particular example, we'll end up with almost exactly the same sample size at the end of our simulations!).

The Test of Proportions is a standard test, and the formula to calculate the corresponding sample size is easily available in R and Python. Let's first look at the R formula.

With an average booking rate of 18.25% in our historical data, the chosen effect size of 1% would translate into an expected booking rate for our treatment group of 19.25%. For the standard values of the parameters—statistical significance = 0.05 and power = 0.8—the corresponding formula in R would be:

```
## R
> effect_size <- ES.h(0.1925,0.1825)
> pwr.2p.test(h = effect_size, n = NULL, sig.level = 0.05, power = 0.8,
                 alternative = "greater")


     Difference of proportion power calculation for binomial distribution
                    (arcsine transformation)

           h = 0.02562255
           n = 18834.47
    sig.level = 0.05
        power = 0.8
  alternative = greater

NOTE: same sample sizes
```

---

4 See Aberson (2019).

The syntax of all the functions for power analysis in the pwr package is the same, with the exception of the notation for the effect size, which changes from one formula to another:

- h is the effect size, based on the increase in probability we want to be able to observe over the baseline probability.
- n is the sample size for each group.
- sig.level is the statistical significance.
- power is the statistical power, equal to $1 - \alpha$.

When entering the formula, you should enter the values for three of these variables and set the remaining one to NULL. In the preceding formula, we're calculating the sample size, so we set n = NULL.

Note that for a test of two proportions, the effect size for statistical purposes depends on the baseline rate; an increase of 5% from a baseline of 10% or 90% is more "important" than from a baseline of 50%. Fortunately, the package pwr provides the ES.h() function, which translates the expected probability and the baseline probability into the right effect size for the formula.

Note also the parameter at the end of the formula: alternative indicates whether you want to run a one-sided (greater or less) or a two-sided (two.sided) test. As long as our treatment doesn't increase our booking rate, we don't really care whether it has the same booking rate or a lower booking rate compared to our control; either way, we won't implement it. This implies that we can run a one-sided test instead of a two-sided test by setting alternative = 'greater'.

The code for Python is similar, using the proportion_effectsize() function from the package statsmodels.stats.proportion:

```
## Python
effect_size = ssprop.proportion_effectsize(0.194, 0.184)
ssp.tt_ind_solve_power(effect_size = effect_size,
                       alpha = 0.05,
                       nobs1 = None,
                       alternative = 'larger',
                       power=0.8)
Out[1]: 18950.818821558503
```

The sample size returned by the formula is 18,800 per group (plus or minus some minor variation between R and Python), i.e., 37,600 total, which means we can achieve the necessary sample size in a bit less than four months. That was easy! Using a statistical significance of 0.1 and a power of 0.9 would yield a sample size of 20,000 per group, just a bit longer.

What does a total sample size of 40,000 for a statistical significance of 0.1 and a power of 0.9 mean in terms of the decision model I outlined in the previous section? Imagine the following:

- You run a very large number of experiments with a total sample size of 40,000, as described.

- Your decision rule in each case is that you will implement the 1-click button if the statistics from the test of proportions has a p-value lower than 0.1.

- In all of these experiments, the true effect size is 1%.

Then you will find a significant positive result and implement the 1-click button in 90% (i.e., 0.9) of these experiments; in the remaining 10% of these experiments, you'll get a null result and wrongly reject implementing the 1-click button.

There are some equivalent formulas for regression, but only for the simplest cases, and I find that even in those situations, their complexity vastly outweighs their usefulness. Nonetheless, as a conceptual step toward our simulation approach, let's review what the traditional statistical approach would look like in terms of the decision model with regression. Let's run a logistic regression on some mock-up data:

```R
## R (output not shown)
exp_null_data <- hist_data %>%
  slice_sample(n=20000) %>%
  mutate(oneclick = ifelse(runif(20000)>0.5,1,0)) %>%
  mutate(oneclick = factor(oneclick, levels=c(0,1)))
summary(glm(booked ~ oneclick + age + gender,
                data = exp_null_data, family = binomial(link = "logit")))
```

```Python
## Python
exp_null_data_df = hist_data_df.copy().sample(2000)
exp_null_data_df['oneclick'] = np.where(np.random.uniform(0,1,2000)>0.5, 1, 0)
mod = smf.logit('booked ~ oneclick + age + gender', data = exp_null_data_df)
mod.fit(disp=0).summary()
```

```
...
                 coef    std err      z     P>|z|   [0.025    0.975]
Intercept       9.5764    0.621   15.412    0.000   8.359    10.794
gender[T.male]  0.1589    0.136    1.167    0.243   -0.108    0.426
oneclick        0.0496    0.136    0.365    0.715   -0.217    0.316
age            -0.3017    0.017  -17.434    0.000   -0.336   -0.268
...
```

The traditional decision rule would be to consider the impact of the 1-click button to be significant and implement it if the corresponding coefficient (here approximately 0.0475) had a p-value less than 0.1. Because it's approximately 0.28 with this mock-up data, we would consider the effect not significant and decline to implement the button (the actual numbers for you will vary randomly depending on your simulation).

Determining the sample size for our analysis based on this approach would entail determining the sample size such that in 90% of a large number of experiments where the true effect is 1%, we would get a p-value for the regression coefficient less than 0.1. But as I described in Chapter 7, this implicitly makes statistical assumptions about our data being normally distributed, which can be problematic, so we'll use Bootstrap simulations instead as we'll now see.

 Using the sample size formula for the test of proportions can still be useful as a quick and fast first step, because its result should be on the same order of magnitude as the final sample size you'll need. A total sample size of 40,000 for the test of proportions means that unless your other predictors have a crazy high predictive power, the order of magnitude for your required sample size is going to be 10,000, and not 1,000 or 100,000 (i.e., your sample size will have five figures). We will start our simulations with a tentative sample size of 20,000, and based on how much effective power it gives us, we'll adjust that number upward or downward.

### Power analysis without statistics: Bootstrap simulations

The traditional statistical analysis made perfect sense when data was limited and calculations were done painstakingly by hand. I strongly believe that it has now outlived its usefulness: Bootstrap simulations offer an alternative that better reflects the realities and needs of applied data analysis. How wrong an experiment can get (e.g., saying that the treatment is 1% better than the control when in reality it's 10% worse) is often a bigger concern for business partners than how likely it is that the difference is zero.[5]

**Connecting simulations and statistical theory.**   When using Bootstrap simulations, our decision rule doesn't rely on p-values. Instead, we implement the treatment if the Bootstrap confidence interval for the coefficient of interest is above a certain threshold, usually zero. If the assumptions of statistical power analysis are verified, Bootstrap simulations yield results that are very similar and intuitively connected:

- Under the sharp null hypothesis of no effect, we expect that a 90%-CI will include zero 90% of the time, an 80%-CI will include zero 80% of the time, and so on. This property, called the *coverage* of the CIs, implies that the percentage we use to define our CI is equivalent to statistical significance, i.e., a 90%-CI will have an approximately 5% false positive rate in each direction. In 5% of cases we'll observe a CI that is strictly negative and in 5% of cases a CI that is strictly positive.

---

5 Hubbard (2010) is a good resource if you want to think more about how to design useful measurements in business, even with very limited information.

- Given an alternative hypothesis, i.e., a target effect size, we can define our power as the percentage of simulations that yield a true positive. For example, if we set the effect of the 1-click button at 1%, simulate a large number of experiments, and observe that 75% of our Bootstrap CIs are strictly positive, then our power is 75%.

 As we'll see in the next chapter, if the assumptions of traditional statistical power analysis are *not* verified, the coverage of the Bootstrap CI may vary. That is, a 90%-CI may include zero more or less than 90% of the time. This effective coverage represents the real risk of false positives that needs to be set to the desired level of significance. This is just a heads-up; we'll get into more details in Chapter 9.

Simulations offer a very versatile but transparent way of determining the necessary sample size for any experiment, however weird the data or complex the business decision at hand. These advantages come from making you state explicitly how you'll analyze your data and write the corresponding code before actually running the experiment, which provides an additional sanity check and opportunity to make adjustments. The counterpart to these benefits is that we'll have to do more coding ourselves instead of relying on off-the-shelf formulas. I'll attempt to limit the complexity of the code by breaking it into intuitive functions.

**Writing our analysis code.** Let's first create a function that will output our metric of interest, namely the coefficient for *OneClick* in our logistic regression:

```
## R
#Metric function
log_reg_fun <- function(dat){
  #Running logistic regression
  log_mod_exp <- glm(booked ~ oneclick + age + gender,
                     data = dat, family = binomial(link = "logit"))
  summ <- summary(log_mod_exp)
  metric <- summ$coefficients['oneclick1', 'Estimate']
  return(metric)}

## Python
def log_reg_fun(dat_df):
    model = smf.logit('booked ~ oneclick + age + gender', data = dat_df)
    res = model.fit(disp=0)
    coeff = res.params['oneclick']
    return coeff
```

This is just a functional wrapper for our preceding analysis, and applying that function to our mock-up data set would return the same coefficient, approximately 0.0475.

Let's then calculate Bootstrap CIs for this metric, reusing the function from Chapter 7:

```Python
## Python
def boot_CI_fun(dat_df, metric_fun, B = 100, conf_level = 0.9):
    #Setting sample size
    N = len(dat_df)
    conf_level = conf_level
    coeffs = []

    for i in range(B):
        sim_data_df = dat_df.sample(n=N, replace = True)
        coeff = metric_fun(sim_data_df)
        coeffs.append(coeff)

    coeffs.sort()
    start_idx = round(B * (1 - conf_level) / 2)
    end_idx = - round(B * (1 - conf_level) / 2)
    confint = [coeffs[start_idx], coeffs[end_idx]]
    return confint
```

With R, we'll directly reuse the `boot_ci()` function I built.

Similarly, we'll take as our decision rule that we'll implement the button if and only if the Bootstrap 90%-CI is strictly positive (i.e., it doesn't include zero):

```R
## R
decision_fun <- function(dat){
  boot_ci_value <- boot_ci(dat, metric_fun)
  decision <- ifelse(boot_ci_value[1]>0,1,0)
  return(decision)}
```

```Python
## Python
def decision_fun(dat_df, metric_fun, B = 100, conf_level = 0.9):
    boot_CI = boot_CI_fun(dat_df, metric_fun, B = B, conf_level = conf_level)
    decision = 1 if boot_CI[0] > 0  else 0
    return decision
```

This is equivalent to the decision rule of implementing the button if and only if the p-value is under the 0.10 threshold. You can check by yourself that applying this function to our mock-up data set returns 0 as it should.

The definition of the power of our experiment for a given effect size and a given sample size remains the same: it is the percentage of a large number of such experiments for which we would implement the button. Let's now turn to simulating this large number of experiments!

**Power simulation.** We'll then write our function to run a single simulation. The code works as follows (the callout numbers apply to both R and Python):

```
## R
> single_sim_fun <- function(dat, metric_fun, Nexp, eff_size, B = 100,
                            conf.level = 0.9){

    #Adding predicted probability of booking ❶
    hist_mod <- glm(booked ~ age + gender + period,
                    family = binomial(link = "logit"), data = dat)
    sim_data <- dat %>%
      mutate(pred_prob_bkg = hist_mod$fitted.values) %>%
      #Filtering down to desired sample size ❷
      slice_sample(n = Nexp) %>%
      #Random assignment of experimental groups ❸
      mutate(oneclick = ifelse(runif(Nexp,0,1) <= 1/2, 0, 1)) %>%
      mutate(oneclick = factor(oneclick, levels=c(0,1))) %>%
      # Adding effect to treatment group ❹
      mutate(pred_prob_bkg = ifelse(oneclick == 1,
                                    pred_prob_bkg + eff_size,
                                    pred_prob_bkg)) %>%
      mutate(booked = ifelse(pred_prob_bkg >= runif(Nexp,0,1),1, 0))

    #Calculate the decision (we want it to be 1) ❺
    decision <- decision_fun(sim_data, metric_fun, B = B,
                            conf.level = conf.level)
return(decision)}

## Python
def single_sim_fun(Nexp, dat_df, metric_fun, eff_size, B = 100,
                  conf_level = 0.9):

    #Adding predicted probability of booking ❶
    hist_model = smf.logit('booked ~ age + gender + period', data = dat_df)
    res = hist_model.fit(disp=0)
    sim_data_df = dat_df.copy()
    sim_data_df['pred_prob_bkg'] = res.predict()
    #Filtering down to desired sample size ❷
    sim_data_df = sim_data_df.sample(Nexp)
    #Random assignment of experimental groups ❸
    sim_data_df['oneclick'] = np.where(np.random.uniform(size=Nexp) <= 0.5, 0, 1)
    # Adding effect to treatment group ❹
    sim_data_df['pred_prob_bkg'] = np.where(sim_data_df.oneclick == 1,
                                            sim_data_df.pred_prob_bkg + eff_size,
                                            sim_data_df.pred_prob_bkg)
    sim_data_df['booked'] = np.where(sim_data_df.pred_prob_bkg >= \
                                    np.random.uniform(size=Nexp), 1, 0)

    #Calculate the decision (we want it to be 1) ❺
    decision = decision_fun(sim_data_df, metric_fun = metric_fun, B = B,
                            conf_level = conf_level)
    return decision
```

**❶** Add the predicted probability of booking to the data.

**❷** Filter down to the desired sample size.

**❸** Assign experimental groups.

**❹** Add effect to the treatment group.

**❺** Apply the decision function and return its output.

We can then write our power function for a certain effect size and sample size. This function repeatedly generates experimental data sets and then applies our decision function to them; it returns the fraction of them for which we would implement the button:

```
## R
power_sim_fun <- function(dat, metric_fun, Nexp, eff_size, Nsim,
                          B = 100, conf.level = 0.9){
  power_list <- vector(mode = "list", length = Nsim)
  for(i in 1:Nsim){
    power_list[[i]] <- single_sim_fun(dat, metric_fun, Nexp, eff_size,
                                      B = B, conf.level = conf.level)}
  power <- mean(unlist(power_list))
  return(power)}

## Python
def power_sim_fun(dat_df, metric_fun, Nexp, eff_size, Nsim, B = 100,
                 conf_level = 0.9):
    power_lst = []
    for i in range(Nsim):
        print("starting simulation number", i, "\n")
        power_lst.append(single_sim_fun(Nexp = Nexp, dat_df = dat_df,
                                        metric_fun = metric_fun,
                                        eff_size = eff_size, B = B,
                                        conf_level = conf_level))
    power = np.mean(power_lst)
    return power
```

How many data sets should you simulate? Twenty is a good starting point; it will give you a noisy estimate, but if you get a power of 0 or 1, you'll know you have to adjust your sample size:

```
## Python (output not shown)
power_sim_fun(dat_df=hist_data_df, metric_fun = log_reg_fun, Nexp = int(4e4),
             eff_size=0.01, Nsim=20)

## R
> set.seed(1234)
> power_sim_fun(dat=hist_data, metric_fun = log_reg_fun, Nexp = 4e4,
+                eff_size=0.01, Nsim=20)
[1] 0.9
```

This first estimate is 90% power; like I said, traditional formulas give you a reasonable ballpark to start your simulations. I then run the power simulation function with 30,000 and 50,000 rows for 100 simulations each, and finally 35,000 and 45,000 rows for 200 simulations each. Basically, as you get a narrower and narrower interval of sample sizes, you want to improve accuracy by increasing the number of simulations. Figure 8-6 shows the results of my successive iterations.

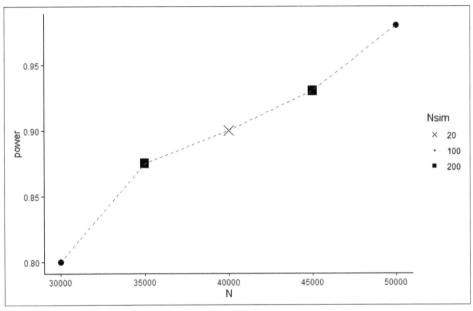

*Figure 8-6. Power simulations for various sample sizes*

As announced earlier, we get a power of 0.9 at about 40,000. We could keep running simulations if we needed to get more precise (e.g., should we get a sample size of 38,000? 41,000?) but that's good enough for this example.

Now that we've determined our sample size, the last thing I like to do is plot the power curve for a few effect sizes at that sample size. This gives us a better sense of how likely overall we are to get a positive result, assuming the actual effect size is positive. It also allows you to better convey to your business partners that the power of your experiment is not just defined for one effect size. Here, we can see how the power of the experiment increases going from a 0.5% effect to a 2% effect (Figure 8-7).

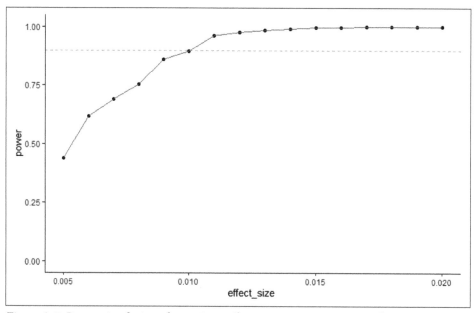

Figure 8-7. Power simulations for various effect sizes at N = 40,000, with 200 simulations per effect size, dashed line at power = 0.9

With 200 simulations per effect size, the estimated values for power should be pretty accurate, although still not perfect, as illustrated by the lack of smoothness of the curve. In other words, the fact that we see a power of 1 for an effect size of 2% doesn't mean that we literally have 100% power for that effect size, but very close to it.

 Reminder: the simulated statistical significance of a Bootstrap CI should be pretty close to a normal CI if your variables are reasonably smoothly and normally distributed. For weirder data (multiple peaks, fat tails, etc.), this may not hold anymore, and you should definitely check that your simulated statistical significance is not widely off.

If you'd like, you can also run a simulation with an effect size of zero. This will give you the empirical statistical significance of your analysis. Because we're using Bootstrap 90%-CIs, about 5% of these simulations should end up with the decision to (wrongly) implement the 1-click button, and this is what we observe here.

These are not big data simulations by any means, but they take long enough (the longest one took about half an hour on my laptop) that you'll want to improve the performance of your functions, get your code running while you do something else, or both. Beside the BehavioralDataAnalysis package (*https://github.com/Behavioral DataAnalysis/R_package*), the code on GitHub (*https://oreil.ly/ BehavioralDataAnalysis*) contains functions that I have optimized using the Rfast and doParallel packages in R and the joblib and psutil packages in Python.

# Analyzing and Interpreting Experimental Results

After running the experiment and collecting the corresponding data, you can analyze them. After all the simulated analyses you ran for the power estimation, the final analyses themselves should be a walk in the park. We run a logistic regression and determine the corresponding Bootstrap 90%-CI. Because of the random assignment, we know that our coefficient for the 1-click button is unconfounded—we don't need to control for any confounder. However, by adding other variables that are also causes of the booking probability, we can reduce noise and significantly improve the precision of our estimation:

```
## Python code (output not shown)
import statsmodels.formula.api as smf
model = smf.logit('booked ~ age + gender + oneclick', data = exp_data_df)
res = model.fit()
res.summary()

## R
> log_mod_exp <- glm(booked ~ oneclick + age + gender,
                     data = exp_data, family = binomial(link = "logit"))
> summary(log_mod_exp)

...
Coefficients:
             Estimate Std. Error z value    Pr(>|z|)
(Intercept)  11.94701    0.22601 52.861     < 2e-16 ***
oneclick1     0.15784    0.04702  3.357     0.000789 ***
age          -0.39406    0.00643 -61.282    < 2e-16 ***
genderfemale -0.25420    0.04905  -5.182 0.000000219 ***
...
```

The coefficient for the 1-click button is equal to 0.15784, and the Bootstrap 90%-CI for it is approximately [0.073; 0.250]. Based on our decision rule, we would go ahead and implement the 1-click button.

The coefficients from a logistic regression are not directly interpretable, and I find that the recommended solution of using the odds ratio only helps marginally (in particular when you have moderation effects). My preferred rule of thumb is to generate two copies of the experimental data, one where the variable for the 1-click button is set to 1 for everyone, and the other where it is set to 0. By comparing the probability of booking predicted by our logistic model for these two data sets, we can calculate an "average" effect that is very close to the effect you would observe if implementing the treatment for everyone. It's unscientific but useful:

```R
## R (output not shown)
> diff_prob_fun <- function(dat, reg_model = log_mod_exp){
    no_button <- dat %>% ❶
      mutate(oneclick = 0) %>%
      mutate(oneclick = factor(oneclick, levels=c(0, 1))) %>%
      select(age, gender, oneclick)
    button <- dat %>% ❷
      mutate(oneclick = 1) %>%
      mutate(oneclick = factor(oneclick, levels=c(0, 1))) %>%
      select(age, gender, oneclick)
    #Adding the predictions of the model
    no_button <- no_button %>% ❸
      mutate(pred_mod = predict(object=reg_model, newdata = no_button,
                                type="response"))
    button <- button %>%
      mutate(pred_mod = predict(object=reg_model, newdata = button,
                                type="response"))
    #Calculating average difference in probabilities
    diff <- button$pred_mod - no_button$pred_mod ❹
    return(mean(diff))}
> diff_prob_fun(exp_data, reg_model = log_mod_exp)
```

```Python
## Python
def diff_prob_fun(dat_df, reg_model = log_mod_exp):

    #Creating new copies of data
    no_button_df = dat_df.loc[:, 'age':'gender'] ❶
    no_button_df.loc[:, 'oneclick'] = 0
    button_df = dat_df.loc[:,'age':'gender'] ❷
    button_df.loc[:, 'oneclick'] = 1

    #Adding the predictions of the model
    no_button_df.loc[:, 'pred_bkg_rate'] = res.predict(no_button_df) ❸
    button_df.loc[:, 'pred_bkg_rate'] = res.predict(button_df)

    diff = button_df.loc[:,'pred_bkg_rate'] \  ❹
    - no_button_df.loc[:,'pred_bkg_rate']
    return diff.mean()
diff_prob_fun(exp_data_df, reg_model = log_mod_exp)
0.007129714313551981
```

❶ We create a data set called no_button for which we set the variable oneclick to zero for all rows (and convert it to factor so that the predict function later will work).

❷ We create a data set called button for which we set the variable oneclick to one for all rows.

❸ We calculate the predicted probability of booking in each case by using the predict() function with our model log_mod_exp.

❹ We calculate the difference between the predicted probabilities.

We can see that our average effect across our experimental population is about 0.712pp, positive but lower than our target of 1pp. As usual, let's build the Bootstrap 90%-CI, which is approximately [0.705pp; 0.721pp]. This interval is very narrow and does not cross zero. Therefore, we can treat our result as empirically statistically significant at the 5% level. In this case, we can even get much more confident than that: the 99.8%-CI is approximately [0.697pp; 0.728pp], still far from zero, so we can treat our result as significant at the (1 − 0.998) / 2 = 0.1% level.

To cover all cases, let's recap our decision rule (Table 8-4). This way, you'll see what to do depending on whether the observed estimated effect is statistically significant or not and economically significant (taken here to mean a 1pp increase) or not. In the present case, I would implement the button, given that it has a strictly positive effect, and the cost of implementation is low.

*Table 8-4. Decision rule for 1-click booking button*

| | | Observed estimated effect | | |
| --- | --- | --- | --- | --- |
| | | estimated effect <= 0 | 0 < estimated effect < 1pp | 1pp <= estimated effect |
| Empirical statistical significance of observed results | "High" (the Bootstrap CI for 90% or higher doesn't cross 0) | Don't implement button | Implement button or not, depending on estimated effect size, costs, and risk appetite (**our case here**) | Implement button |
| | "Low" (the Bootstrap CI for 90% crosses 0) | Don't implement button | Don't implement button | Implement button or run new test, depending on confidence interval and risk appetite |

The last thing I will note is that the average effect across our experimental population, 0.712pp, is pretty far from the straightforward difference between our control group and treatment group, which is about 0.337pp. This is due to random differences between our two experimental groups. The average age in our control group is 40.63 years versus 40.78 in the treatment group. The proportion of men is also a bit higher in the treatment group. With very small effect sizes, these minute differences are enough to muddy a direct comparison of the two groups: our sample size is large enough that our two groups are identical within about 0.3pp, which is pretty close in absolute terms, but that's about half of our experimental effect.

Unfortunately, there's nothing we can do about that in this experiment, where customers are allocated randomly to the two groups as they come. But if we know our entire experimental sample at the beginning of the experiment, we can do significantly better by ensuring that the control group and the experimental group are as identical as possible through stratified randomization, as we'll see in the next chapter.

## Conclusion

In this chapter, we saw how to design the simplest form of experiment, an online A/B test with straightforward randomization. I emphasized that a well-designed experiment is much more than just throwing out random different versions of a website or an email to customers. You need to determine your business goal and target metric, and then articulate how your intervention is connected to them through behavioral logic. Taken together, your business goal, target metric, intervention, and behavioral logic constitute the theory of change of your experiment.

We then turned to the quantitative aspects of experimental design. In this first chapter on experimentation, the random assignment was extremely simple, and I spent more time on the power analysis and sample size calculations. While there are statistical formulas available, I prefer to use regressions rather than statistical tests as analysis tools, and Bootstrap confidence intervals rather than p-values as a measure of the uncertainty around our estimated coefficient, which leads to using power simulations instead of formulas. In this case, the results of the two are almost identical, but in the next two chapters we'll get into more complex designs where there are no formulas available.

# CHAPTER 9

# Stratified Randomization

In the previous chapter, we saw the simplest form of randomization: a customer shows up, and we toss a metaphorical coin or dice. Heads and they see version A, tails and they see version B. The probabilities may be different from 50/50, but they are constant, and independent of the customer characteristics. No "my control group is a bit older than my treatment group, let's make sure the next Millennial who shows up goes into the control group." As a consequence, your control and treatment groups are "probabilistically equivalent," which is statistics' way of saying that if you kept running your experiment forever, your two groups would have the exact same proportions as your general population. In practice, however, your experimental groups; can end up being quite different from each other. Adding explanatory variables to your final analysis can somewhat compensate for these imbalances, but as we'll now see, we can do better than that if we know ahead of time who is going to be part of our experiment.

In this chapter, I'll introduce you to stratified randomization, which will allow us to ensure that our experimental groups are as similar as possible. This starkly increases the explanatory power of an experiment, which is especially useful when you can't have large sample sizes.

Stratified randomization can be applied to any situation where we have a predetermined list of customers/employees/etc. to build our experimental groups from. Given that A/B tests are most often discussed in relation to minor changes to an email or website, I could have taken the example of an email campaign. But I wanted to demonstrate that larger business initiatives, which are often taken by company executives based on their "strategic sense," can also be tested and validated.

The business context here is that by default, AirCnC gives owners at least 24 hours to clean their property between two bookings. In high-demand markets where properties are booked as soon as they're available, this represents a significant limiting factor. Business leaders are keen to reduce it and increase monthly profit per property. As is often the case, there are two schools of thought in the company:

- The finance department advocates offering owners the opportunity to set a minimum duration of two nights per booking.
- The customer experience department believes a minimum duration would adversely impact customer satisfaction; instead, it advocates offering owners the services of a professional cleaning company for free in exchange for a reduction of the cleaning window from 24 to 8 hours.

Situations like this often arise in business. Both sides have somewhat compelling arguments that speak to different aspects of the problem or emphasize different metrics (here, booking profit versus customer experience), and/or provide anecdotal evidence in favor of their position ("this other company does X, so we should do it too"). A common outcome is that whoever has the most sway "wins" and their solution gets implemented, a.k.a. organizational politics rule.

At this point, you probably expect me to say something along the lines of "but experimentation allows you to bypass all politics and reach the best solution without any fuss." I wish it were that easy! The truth is that experimentation can be of tremendous help in such situations, but it's not a cure-all, for two reasons.

The first reason is that unless a solution turns out to be superior to the other on all fronts, there will be some genuine trade-offs that need to be made between competing objectives: how much decrease in customer satisfaction is the company willing to tolerate for an increase in profit? That question is inherently political because different stakeholders in the company have different preferences on that matter. You'll have to present these trade-offs as clearly as possible to your leaders and get them resolved as much as possible ahead of time if you want your experiment to be successful.

The second reason is that your experiment will make at most one side happy (it can also make both sides unhappy if the control group has the best outcomes!). Unhappy business leaders, like any other unhappy human being, can be very good at finding rationalizations after the fact: "The San Francisco Bay area is different from other high-demand markets," "Survey-measured customer satisfaction is down, but Net Promoter Score is up, and NPS is a better measure of 'true' customer satisfaction," etc.

These two reasons make it extra important to properly plan and run your experiment, not just from an experimental design perspective, but also from a business perspective.

# Planning the Experiment

As we've seen in the previous chapter, successfully planning an experiment requires us to clearly articulate its theory of change:

- What are the business goal and the target metric?
- What is the definition of our intervention?
- How are these connected by our behavioral logic?

You should hopefully by now be familiar with that process, so I will not elaborate on how to do it, and I'll just go through the steps quickly to give you a deeper understanding of the experiment. In particular, I'll use the opportunity to call out some peculiarities of the experiment from a behavioral perspective.

## Business Goal and Target Metric

The business goal for this experiment, or equivalently the business problem we're trying to solve, is to increase profitability by reducing the amount of downtime in high-demand markets. Because the cost of offering free cleaning services to owners is significant (finance has estimated it to be $10/day), we'll need to include it in our analysis.

We'll do so by modifying our target metric accordingly. Our basic metric is average booking profit per day; we'll use instead average booking profit per day *net of extra cost*. This simply means that we'll have to deduce $10 from the basic metric for the free cleaning treatment.

However, there are also some concerns that the minimum-duration intervention would negatively impact customer satisfaction. How can we take that into account?

A solution I have seen other authors advocate is to use a weighted average of metrics, (sometimes called an Overall Evaluation Criterion [OEC]). In our present example, this would mean assigning weights, e.g., 50% each, to our two variables, and then using that new metric as our target. You should certainly feel free to use that approach if you or your business partners wish to, but I would recommend instead picking a unique target metric, and if necessary have a few other "guardrail" metrics on a watch list, for several reasons.

The first one is that the trade-offs between business goals are ultimately strategic and political decisions. There is no objective best answer as to how many CSAT points an increase in profit is worth; it depends on the organization, its context, and its current priorities. Using a weighted average determined at one point in time gives the process the appearance of technical objectivity but it's really only fossilized subjectivity.

The second is that an OEC makes these trade-offs linear. If one CSAT point is equivalent to $10 million of profit with an OEC, then five CSAT points are equivalent to $50 million of profit. But a decrease of one CSAT point might mean fewer delighted customers while a decrease of five points might mean a social media storm. A dollar is always worth another dollar, but for pretty much anything else, a series of small changes is generally preferred to a single big one. Relying blindly on an OEC can lead to riskier decisions. Proponents of the OEC approach may object at this point that obviously you shouldn't rely on it blindly; but if you're going to look at the various components and have an open discussion anyway, I'm not quite sure how having an OEC helps.

Moreover, an OEC takes business interventions as fixed. Sticking with the 1 point CSAT = $10 million equivalence, the following two options would have an equal rating of zero:

1. The first intervention increases profit by $1 million and decreases CSAT by 0.1 points.

2. The second intervention increases profit by $50 million and decreases CSAT by 5 points.

But there is a big difference from a behavioral perspective. The first option is basically a dead horse and there is little hope of beating life into it, whereas the second option is more like a mercurial purebred. By targeting it to a specific segment of customers, changing its terms or the way it's presented, there might be a way to reap at least some of its benefits without its costs. In that sense, an intervention with both big positive and negative effects calls for exploration and design iterations rather than the binary decisions encouraged by an OEC.

Finally, I believe that in some cases, an OEC is used as a shortcut. Let's say that an intervention increases short-term profit but also increases the probability of defection. This is not a genuine strategic trade-off: we should measure the impact of the probability of defection on lifetime value, and then determine the net effect on profitability. Saying that your OEC will be 90% short-term profit and 10% effect on defection rate is a way of taking a guess instead of measuring the true exchange rate. With the causal-behavioral framework of this book, we can do better than taking guesses.

Therefore, for the rest of the chapter, we'll use average booking profit per day as our single target metric, and assume that CSAT is being monitored in the background for any worrying change.

That's all well and good, but when we're talking about tracking customer satisfaction, which customers are we talking about? If customers want to book a location for just one night and are offered a location with a duration minimum, they might decide to book a different one (either in the control group or the free-cleaning group), or else completely give up on booking through AirCnC and book a hotel instead. Therefore, we can't simply measure customer experience for a clearly defined group of customers.

Unfortunately, that problem is not rare: whenever you run an experiment where the experimental unit for random assignment is not a customer, you have to ask yourself how this will play out on the customer side. What we can do is leverage the fact that minimum duration would impact only customers who are looking for a one-night booking; we also know that customers enter their desired duration before being shown available properties. Therefore, we could track all customers in our experiment who have a one-night desired duration and check whether they end up booking several nights in the same property or one night in another property, or end up not booking at all. Whenever they make a booking, we would also track how they rate their stay. That's certainly far from perfect because we would be tracking different metrics for different subgroups of customers, but that's the best we can do, and is another example of why I believe much more in monitoring guardrail variables than in aggregating them in an OEC.

## Definition of the Intervention

After determining the criteria for success, we need to make sure we have clarity on what we are testing. This is especially important when the organizational stakes are high, e.g., business leaders are butting heads on the question. Here, it is important to point out that we're offering owners the opportunity to set a minimum duration, which is not the same thing as making a minimum duration mandatory. Similarly, owners may or may not take the free-cleaning offer. In addition, the treatments themselves could still be open to interpretation. How thorough and costly for the company is the free cleaning? What is the minimum duration we're enforcing on customers?

Because both of our interventions are somewhat complex and rely on owners understanding an offer and choosing to take it, it is probably a good idea to create a few different designs and test them qualitatively through UX research. In addition, after running the experiment, it would be a good idea to consider slightly different variations of whichever treatment we decide to implement.

Ultimately, you'll want to make sure that all the stakeholders are satisfied with the designs for the experiment and are willing to sign off on it. It will reduce (but not eliminate!) the risk that when the results come in, they will argue that what was tested did not adequately represent their proposed solution.

## Behavioral Logic

The behavioral logic is different for the two treatments: the minimum-duration approach would increase the duration and amount per booking but may reduce the overall number of bookings; on the other hand, the free-cleaning approach would increase the number of bookings but will reduce the profit per booking due to the added cost. In addition, we need to take into account that our experimental treatments/interventions are *offers*, which owners may or may not accept (Figure 9-1).

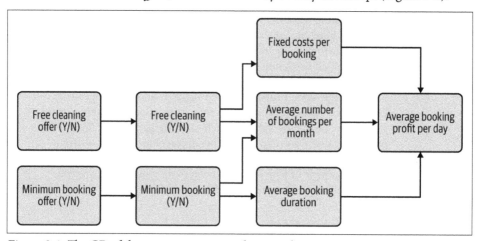

*Figure 9-1. The CD of the two treatments under consideration*

## Data and Packages

The GitHub folder for this chapter (*https://oreil.ly/BehavioralDataAnalysisCh9*) contains two CSV files with the variables listed in Table 9-1. The check mark (✓) indicates the variables present in that file, while the cross (✕) indicates the variables that are not present.

*Table 9-1. Variables in our data*

|  | Variable description | chap9-historical_data.csv | chap9-experimental_data.csv |
|---|---|---|---|
| ID | Property/owner ID, 1-5000 | ✓ | ✓ |
| sq_ft | Square footage of property, 460-1120 | ✓ | ✓ |
| tier | Tier of property, categorical, from 1 to 3 in descending tiers | ✓ | ✓ |
| avg_review | Average review of property, 0-10 | ✓ | ✓ |
| BPday | Booking profit per day, target variable, 0-126 | ✓ | ✓ |

| | Variable description | chap9-historical_data.csv | chap9-experimental_data.csv |
|---|---|---|---|
| *Period* | Month index, 1-35 in historical data, 36 implicit in experimental data | ✓ | |
| *Month* | Month of year, 1-12 | ✓ | ✓ |
| *grp* | Experimental assignment, "ctrl," "treat1" (free cleaning), "treat2" (min. booking duration) | ✗ | ✓ |
| *compliant* | Binary variable indicating if owner was treated according to their assigned group | ✗ | ✓ |

For this chapter, R users will be able to leverage dedicated functions from the `Behav ioralDataAnalysis` package, which will make their life easier.

Python users will use instead some custom code, as well as functions from the following packages in addition to the common ones:

```
## Python
import random # For functions sample() and shuffle()
# To rescale numeric variables
from sklearn.preprocessing import MinMaxScaler
# To one-hot encode cat. variables
from sklearn.preprocessing import OneHotEncoder
```

# Determining Random Assignment and Sample Size/Power

For this experiment, we'll assign experimental groups all at once, based on the list of our owners at a point in time. This gives us an opportunity to significantly improve over purely random allocation by ensuring from the get-go that our two groups are well balanced through a method called stratification. This will allow us to get more statistical power out of any given sample size.

Therefore, I'll first explain the method for the random assignment, so that we can use it in the simulations for our power analysis. Finally, we'll compare the results of these simulations with a traditional statistical power analysis.

## Random Assignment

Before getting into stratification, let's see what standard randomization would look like.

### Random assignment level

The first consideration for our random assignment is the level at which we'll implement it and then measure the outcomes of the experiment. In the previous chapter, I discussed the question of whether the random assignment should happen at the level of a customer or of a booking. In the present case, the logistics of the experimental treatments, especially the free-cleaning one, precludes implementation at the booking level. Therefore, we'll do the random assignment at the level of a property owner, which number 5,000 in AirCnC's data at this point in time.

## Standard randomization

The process is similar to the one we used in the previous experiment, but simpler and more accurate because it can be done offline instead of in real time: as we already know our population size, we can split it exactly between our experimental groups. We do this by generating an equal number of the values for each of the experimental groups and shuffling them across individuals. The following code illustrates the approach with three groups and a sample size of 999 (to demonstrate the sampling down):

```R
## R
no_strat_assgnt_fun <- function(dat, Nexp){
  dat <- dat %>%
    distinct(ID) %>%
    slice_sample(n=Nexp)
  grp_vec <- c("ctrl", "treat1", "treat2") %>% rep(Nexp/3) %>% sample()
  dat <- dat %>%
    mutate(grp = grp_vec) %>%
    mutate(grp = factor(grp)) %>%
  return(dat)
}
no_strat_assgnt_data <- no_strat_assgnt_fun(hist_data, Nexp = 999)
```

```Python
## Python
def no_strat_assgnt_fun(dat_df, Nexp):
    temp_df = pd.DataFrame({'ID': hist_data_df.ID.unique()})
    temp_df = temp_df.sample(Nexp)
    grp_lst = ['ctrl', 'treat1', 'treat2'] * int(Nexp / 3)
    random.shuffle(grp_lst)
    temp_df['grp'] = grp_lst
    return temp_df
no_strat_assgnt = no_strat_assgnt_fun(hist_data_df, Nexp = 999)
```

One nice aspect of this approach is that it can easily be generalized to an arbitrarily large number of groups; the control group is then tagged with 0, the first treatment group 1, and so on:

```R
## R
no_strat_assgnt_K_fun <- function(dat, Nexp, K){
  dat <- dat %>%
    distinct(ID) %>%
    slice_sample(n=Nexp)
  grp_vec <- seq(K) %>% rep(Nexp/K) %>% sample()
  dat <- dat %>%
    mutate(grp = grp_vec) %>%
    mutate(grp = factor(grp))
  return(dat)
```

```
    }
no_strat_assgnt_data <- no_strat_assgnt_K_fun(hist_data, Nexp = 2000, K = 4)

## Python
def no_strat_assgnt_K_fun(dat_df, Nexp, K):
    temp_df = pd.DataFrame({'ID': hist_data_df.ID.unique()})
    temp_df = temp_df.sample(Nexp)
    grp_lst = list(range(K)) * int(Nexp / K)
    random.shuffle(grp_lst)
    temp_df['grp'] = grp_lst
    return temp_df
no_strat_assgnt = no_strat_assgnt_K_fun(hist_data_df, Nexp = 2000, K = 4)
```

However, an issue with the previous approach is that experimental groups are unlikely to be perfectly balanced across customer characteristics. In order to create balanced experimental groups, we'll want to use a technique called stratification.

## Stratified randomization

Why is a purely random allocation not our best choice? Let's imagine that we are running an experiment on 20 customers, 10 of which are male and 10 of which are female. If we randomly assign each customer to either the control or the treatment group with probability 50%, we expect that on average, there will be 5 males and 5 females in each of the two groups. "On average" here means that if we repeated this assignment a large number of times, the average number of males in the control group across assignments would be 5. But in any given experiment, there's only a 34.4% chance that we get exactly 5 males and 5 females in each group, and there's a 8.9% chance that we get 7 or more men in one group, based on the hypergeometric distribution.[1] Obviously, this problem gets less pronounced with larger sample sizes. With 100 males and 100 females, the probability of getting 70 men or more in one group becomes negligible. But we don't care only about gender: ideally, we would also want a good balance of age, state of residence, usage pattern, etc. This would ensure that our results are relevant for our entire customer base as much as possible, and not just to a specific subgroup of it.

Fortunately, when we have the luxury of assigning experimental groups to all individuals at once, for instance, we can do significantly better than just crossing our fingers and hoping for the best. We can stratify our data: we create "layers" of similar customers, called strata,[2] and we split them between our experimental groups. In the case of our 10 male and 10 female customers, we would create a layer of men, 5 of which would go to the control group and 5 of which to the treatment group, and similarly for women. This implies that each individual still has a 50% chance of ending in

---

[1] Thanks to Andreas Kaltenbrunner for pointing out that this is a hypergeometric and not a binomial distribution.

[2] This is the plural of the Latin word for layer, *stratum*, hence the word stratification.

either group but our control and treatment groups are now perfectly balanced for gender.

Stratification can be applied to any number of variables. With gender and state of residence, we would create a stratum of all women from Kansas and split it equally between our control and treatment group, and so on. With a large number of variables, or with continuous variables, it becomes impossible to find exact matches; in our data, we may not have two women in Kansas with the exact same age and whose properties have the exact same square footage. The solution is to create pairs of individuals that are "as similar as possible," e.g., a 58-year-old woman with a 900-square-foot property and a 56-year-old woman with a 930-square-foot property, and then assign at random one of them to the control group and the other to the treatment group. This way, they still have the same probability individually of ending in any experimental group. When we have only two individuals per strata, this is also called "matching," because we're creating matching pairs of customers.

As often, the intuition is clear enough, but the devil is in the detail of the implementation. There are two steps here:

1. Give a mathematical meaning to the phrase "as similar as possible."
2. Efficiently go through our data to allocate each customer to a pair.

The mathematical concept that we'll use to express "as similar as possible" is distance. Distance can be applied easily to a single numeric variable. If one owner is 56 years old and another is 58 years old, the distance between them is 58 − 56 = 2 years. Similarly, we could say that the distance between a 900-square-foot property and a 930-square-foot property is 30 square feet.

The first complication comes in aggregating multiple numeric variables. We could simply add (or equivalently, take the mean of) the two numbers and say that our two owners are "distant" by 2 + 30 = 32 distance units. The problem with that approach is that square-footage numbers are much bigger than age numbers, as we can see in our example. A difference of 30 years between two owners is likely to be much more important behaviorally than a difference of 30 square feet between their properties. This can be resolved by rescaling all our numeric variables so that their minimum is reset to 0 and their maximum to 1. This means that the "distance" between the youngest and the oldest owners is 1, and the distance between the smallest and the biggest property is also 1. This is not a perfect solution, especially when you have outliers, but it's fast and good enough for most purposes.

The second complication comes from categorical variables. What is the "distance" between a townhouse and an apartment? Or between having a pool or not? A common solution is to say that the distance between two properties is 0 if they are in the same category and 1 otherwise. For example, a townhouse and a house would have a distance of 1 for the property type variable. Mathematically, this is done by *one-hot encoding* categorical variables: that is, we create as many binary 0/1 variables as we have categories. For example, we would transform property type = ("house," "townhouse," "apartment") into three variables, *type.house, type.townhouse,* and *type.apartment.* A property that is an apartment would have a value of 1 for the variable *type.apartment* and 0 for the other two variables. This also has the added advantage of making categorical "distances" comparable with numerical distance. In effect, we're saying that the difference between a townhouse and an apartment is as important as the difference between the smallest and the largest properties. This is again debatable from a behavioral perspective, but that's a good starting point, and often a good stopping point as well.

Let's turn to code, starting with Python. The first step is to aggregate the data at the property level, rescale numeric variables, and one-hot encode categorical variables. This is just boilerplate code, so just skip that bit of code if you don't care about the details of the implementation:

```
## Python code (output not shown)
def strat_prep_fun(dat_df):
    # Making a copy of the input data to avoid side effects
    temp_df = dat_df.copy()

    # Extracting property-level variables
    temp_df = temp_df.groupby(['ID', 'tier']).agg(
        sq_ft = ('sq_ft', 'mean'),
        avg_review = ('avg_review', 'mean'),
        BPday = ('BPday', 'mean'))
    temp_df = temp_df.dropna().reset_index()

    num_df = temp_df.copy().loc[:,temp_df.dtypes=='float64'] #Numeric vars
    cat_df = temp_df.copy().loc[:,temp_df.dtypes=='category'] #Categorical vars

    #Normalizing all numeric variables to [0,1]
    scaler = MinMaxScaler()
    scaler.fit(num_df)
    num_np = scaler.transform(num_df)

    #One-hot encoding all categorical variables
    enc = OneHotEncoder(handle_unknown='ignore')
    enc.fit(cat_df)
    cat_np = enc.transform(cat_df).toarray()

    #Binding arrays
    data_np = np.concatenate((num_np, cat_np), axis=1)
    del num_df, num_np, cat_df, cat_np, enc, scaler
    return data_np

prepped_data_np = strat_prep_fun(hist_data_df)
```

Note that 5,000 is not divisible by 3, so we also need to drop at random two rows, the closest smaller number divisible by 3 being 4,998. Once we have prepared our data, the second step is to create pairs. This computationally intensive problem quickly becomes intractable with larger data (at least if you want the optimal solution). Fortunately, algorithms have been created that will handle it for you, but they will still require some wrapper code.

I wrote a function, `stratified_assgnt_fun()`, which does the job for simple cases; you can find it in the book's GitHub repo (*https://oreil.ly/BehavioralDataAnalysis*). It takes as arguments a dataframe of information on the subject base, the number of subjects required for the experiment, and the number of experimental groups required (including control, i.e., 2 for a standard A/B test):

```
## Python
#Sampling a random monthly period
per = random.sample(range(35), 1)[0] + 1
sample_df = hist_data_df.loc[hist_data_df.period == per].sample(5000)
stratified_data_df = stratified_assgnt_fun(sample_df, K=3)
```

This function takes about 30 seconds to run on my laptop for 5,000 subjects. I also created a simple function to compare the results of the stratified assignment with a purely random assignment, `assgnt_comparison_fun()`, which takes as arguments the dataframe generated by the previous function and the name of a numeric variable we want to use for comparison and returns the standard deviation (s.d.):

```
## Python
>> assgnt_comparison_fun(sample_df, 'sq_ft')
the s.d. between grps for sq_ft is 1.8112  for stratified assignment
the s.d. between grps for sq_ft is 2.3705 for non-stratified assignment

assgnt_comparison_fun(sample_df, 'BPday')
the s.d. between grps for BPday is 0.3305  for stratified assignment
the s.d. between grps for BPday is 0.3674 for non-stratified assignment
```

As you can see, even a very simple function hacked together in a few hours can make your experimental groups much more similar than a purely random assignment.

In R, I have productionalized the stratification code within the `BehavioralDataAnaly sis` package. The only preparatory work required is to aggregate the historical data to the property level, so that we have one row per property.

```
## R
> # Aggregating data to property level
> property_data <- hist_data %>%
+     # We group by all the variables that are already constant
+     group_by(ID, tier, sq_ft, avg_review) %>%
+     # And summarize the others by taking the average value
+     summarize(BPday = mean(BPday)) %>%
+     ungroup()
>
```

```
> head(property_data,5)
# A tibble: 5 × 5
   ID    tier  sq_ft avg_review BPday
  <chr> <fct> <dbl>      <dbl> <dbl>
1 1      2    822.       9.39  45.9
2 10     2    978.      10     47.0
3 100    1    772.       5.05  36.0
4 1000   3    896.       8.99  41.2
5 1001   1    818.       8.14  44.8
```

Then the function paired_assign() from the BehavioralDataAnalysis package takes care of the stratification for you under the hood. After that, we'll just rename the treatment groups for convenience.

```
## R
> stratified_data <- paired_assign(property_data, id = "ID", n.groups = 3)
> stratified_data <- stratified_data %>%
+    mutate(grp = factor(grp)) %>%
+    mutate(grp = fct_recode(grp, ctrl = "0", treat1 = "1", treat2 = "2"))
> head(stratified_data)
    ID tier    sq_ft avg_review   BPday    grp
1    1    2 821.6755   9.393427 45.93412 treat2
2   10    2 977.6863  10.000000 47.02031 treat1
3  100    1 772.2464   5.053913 36.03493   ctrl
4 1000    3 895.7686   8.994171 41.23598   ctrl
5 1001    1 817.6614   8.141247 44.82962 treat2
6 1002    3 968.9046   8.427153 44.36563 treat1
```

The function paired_assign() takes the individual-level data as its first argument, Tidyverse-style.

The parameters for that function are:

- id, the variable(s) used to identify the individuals in the data.
- n.groups, the number of experimental groups, including the control group.

Note that the function automatically dropped 2 properties to create groups of the same size.

Beyond helping reduce noise in our experiments, stratification is also helpful if you intend to do subgroup or moderation analysis (which we'll discuss later in the book). As the phrase goes, "block [with stratification] what you can and randomize what you cannot" (Gerber and Green, 2012).

Stratified randomization is an effective and robust approach to experimental allocation. Its robustness comes largely from its transparency: you can always check afterward that your experimental groups are well-balanced in terms of the means of numeric variables and the proportions of categories for categorical variables.

In addition, because each individual in a pair has the same probability of ending up in any experimental group, even a poorly or wrongly defined distance function will leave you no worse than a purely random allocation. The main risk is to include too many irrelevant variables, which then drown out the relevant variables. That is however easily remedied by including only variables that are part of your causal diagram or the main demographic variables. Don't add a load of other variables simply because you can.

Categorical variables with a large number of categories can also sometimes add noise to your stratification because of their coarseness. Taking the example of employment, a data scientist is different from a statistician, but intuitively, that difference is less than their joint difference with, say, a firefighter. Very granular variables ignore such nuances, and are better replaced with broader categories.

Lest these caveats discourage you: stratification is effective and robust; don't be afraid to use it. Even stratifying based only on a few key demographic variables leads to significant improvements and should be your default approach.

Now that we've determined how we're going to do the random allocation, let's do our power analysis to determine the sample size.

## Power Analysis with Bootstrap Simulations

After concertation with the business partners, we determine that we want to have 90% power for an increase in net booking profit (BP) per day of $2, as it is the minimal observable effect they would be interested in. This translates into an increase in "raw" BP/day of $12 for the free-cleaning intervention (treatment 1) and $2 for the minimum-duration intervention (treatment 2). This won't impact our analyses in any material way, as we can just shift the outcome variable by deducing the $10 cost from the BP/day for properties in the free-cleaning group, but we'll need to keep that in mind and remember to do it. For the sake of simplicity, I'll only discuss the minimum-duration intervention in our power analyses.

Simulation methods really shine in situations such as this one, where there are often no available dedicated formulas to calculate power or sample sizes, or available formulas get horribly complex. The alternative would be to use standard formulas that disregard the specifics of the situation (here, the stratification of our experimental data) and cross our fingers that things go well (Murphy's law: they probably won't).

Our process will be the same as in Chapter 8:

1. First, we'll define our metric function and our decision function.

2. Then we'll create a function that simulates a single experiment for a given sample size and effect size.

3. Finally, we'll create a function that simulates a large number of experiments and count how many of them result in a true positive (i.e., our decision function adequately captures the effect); the percentage of true positives is our power for that sample size.

### Single simulation

Our metric function for the minimum-duration treatment is as follows in Python:

```
## Python
  def treat2_metric_fun(dat_df):
      model = ols("BPday~sq_ft+tier+avg_review+grp", data=dat_df)
      res = model.fit(disp=0)
      coeff = res.params['grp[T.treat2]']
      return coeff
```

In R, we have two options: the first is to create a metric function that returns a single value, similar to the Python one. The second is to directly pass the regression formula to the boot_ci() function but then it will return confidence intervals for all the coefficients in the regression and we'll have to change the decision function accordingly. I'll use the first option, to make things closer to the Python implementation, and make the output of boot_ci() more manageable.

```
## R
> # Metric function for minimum booking duration (treatment 2)
> treat2_metric_fun <- function(dat){
+     lin_model <- lm(BPday~sq_ft+tier+avg_review+grp, data = dat)
+     summ <- summary(lin_model)
+     coeff <- summ$coefficients['grptreat2', 'Estimate']
+     return(coeff)}
>
> set.seed(1)
> treat2_metric_fun(stratified_data)
[1] 0.01936045
```

The metric function for treatment 1 is defined similarly.

We'll reuse the functions for the Bootstrap confidence intervals and decision criterion from Chapter 8. In other words, our decision rule will be to implement the treatment if its 90%-CI is strictly above zero. I'm repeating their code in the following, just for reference:

```
## R
decision_fun <- function(dat, metric_fun, B = 100, conf.level = 0.9){
  boot_ci_values <- boot_ci(dat, metric_fun, B = B, conf.level = conf.level)
  decision <- ifelse(boot_ci_values[1]>0,1,0)
  return(decision)
}

## Python
def boot_CI_fun(dat_df, metric_fun, B = 100, conf_level = 0.9):
  #Setting sample size
  N = len(dat_df)
  coeffs = []

  for i in range(B):
      sim_data_df = dat_df.sample(n=N, replace = True)
      coeff = metric_fun(sim_data_df)
      coeffs.append(coeff)

  coeffs.sort()
  start_idx = round(B * (1 - conf_level) / 2)
  end_idx = - round(B * (1 - conf_level) / 2)
  confint = [coeffs[start_idx], coeffs[end_idx]]
  return confint

def decision_fun(dat_df, metric_fun, B = 100, conf_level = 0.9):
    boot_CI = boot_CI_fun(dat_df, metric_fun, B = B, conf_level = conf_level)
    decision = 1 if boot_CI[0] > 0  else 0
    return decision
```

We can then write the function to run a single simulation, which embeds the logic we've seen so far:

```
## R
single_sim_fun <- function(dat, metric_fun, Nexp, eff_size, B = 100,
                           conf.level = 0.9){
  #Filter the data down to a random month ❶
  per <- sample(1:35, size=1)
  sample_month_dat <- dat %>%
    filter(period == per)

  #Prepare the stratified assignment for a random sample of desired size ❷
  # from the property-level data
  stratified_assgnt <- property_data %>%
    # Reducing the sample size if appropriate
    slice_sample(n=Nexp) %>%
    #Stratified assignment
    paired_assign(id = "ID", n.groups = 3) %>%
    mutate(grp = factor(grp)) %>%
    mutate(grp = fct_recode(grp, ctrl = "0", treat1 = "1", treat2 = "2")) %>%
    #extract the ID and group assignment
    select(ID, grp)
```

```
    sim_dat <- sample_month_dat %>%
      #Apply assignment to full data ❸
      inner_join(stratified_assgnt, by="ID") %>%
      #Add target effect size
      mutate(BPday = ifelse(grp == 'treat2', BPday + eff_size, BPday))

   #Calculate the decision (we want it to be 1 for a positive effect size) ❹
   decision <- decision_fun(sim_dat, metric_fun, B = B, conf.level = conf.level)
   return(decision)
}

## Python
def single_sim_fun(dat_df, metric_fun, Nexp, eff_size, B = 100,
                   conf_level = 0.9):

    #Filter the data down to a random month ❶
    per = random.sample(range(35), 1)[0] + 1
    dat_df = dat_df.loc[dat_df.period == per]
    dat_df = dat_df.sample(n=Nexp)

    #Prepare the stratified assignment for a random sample of desired size ❷
    assgnt = strat_assgnt_fun(dat_df, Nexp = Nexp)
    sim_data_df = dat_df.merge(assgnt, on='ID', how='inner')

    #Add target effect size ❸
    sim_data_df.BPday = np.where(sim_data_df.grp == 'treat2',
                                 sim_data_df.BPday + eff_size, sim_data_df.BPday)

    #Calculate the decision (we want it to be 1) ❹
    decision = decision_fun(sim_data_df, metric_fun, B = B,
                            conf_level = conf_level)
    return decision
```

❶  Select a month at random, to mimic the way we will run our actual experiment (we don't want to use data from 10 years apart in our power analysis).

❷  Generate stratified random assignment for a sample of desired size.

❸  Apply the assignment to the data and apply the target effect size to treatment group 2.

❹  Apply the decision function and return its output.

## Simulations at scale

From there, we can apply the same overall function for power simulations as in Chapter 8 (repeated in the following for reference):

```
## R
> #Standard function for simulations at scale
> power_sim_fun <- function(dat, metric_fun, Nexp, eff_size, Nsim, B = 100,
+                            conf.level = 0.9){
+   power_list <- vector(mode = "list", length = Nsim)
+   for(i in 1:Nsim){
+     power_list[[i]] <- single_sim_fun(dat, metric_fun, Nexp, eff_size, B = B,
+                            conf.level = conf.level)}
+   power <- mean(unlist(power_list))
+   return(power)
+ }
> power_sim_fun(hist_data, treat2_metric_fun, Nexp = 5000, eff_size = 2,
+               Nsim = 100, B = 100, conf.level = 0.9)
[1] 0.98

## Python
def power_sim_fun(dat_df, metric_fun, Nexp, eff_size, Nsim, B = 100,
                  conf_level = 0.9):
    power_lst = []
    for i in range(Nsim):
        power_lst.append(single_sim_fun(dat_df, metric_fun = metric_fun,
                                        Nexp = Nexp, eff_size = eff_size,
                                        B = B, conf_level = conf_level))
    power = np.mean(power_lst)
    return power
```

Our maximum sample size would be 5,000, because that's the total number of property owners AirCnC has. That's a manageable number for simulation purposes, so let's first run 100 simulations with that sample size, which takes 17 minutes on my laptop. There's no point in working our way up to that if it turns out that we would need to use our entire population for the experiment. We find a power of almost 1, which is comforting: the required sample size is less than our total population. From there, we try different sample sizes, progressively increasing the number of simulations as we zero in on the sample size with a power of 0.90 (Figure 9-2).

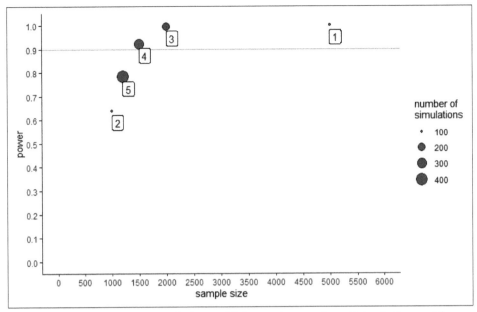

Figure 9-2. Iterative power simulations with increasing number of simulations, with labels indicating the order of runs

It looks like a sample size of 1,500 will do. As in Chapter 8, let's now determine the power curve for various effect sizes at that sample size (Figure 9-3).

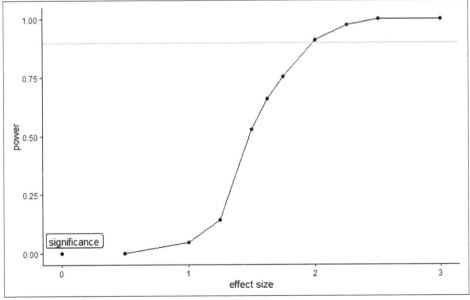

Figure 9-3. Power to detect various effect sizes and significance (sample size = 1,500)

As you can see in Figure 9-3, our power curve drops very steeply, going from an effect size of $2 to $1, and our power is almost nil for effect sizes less than 1; that is, if we assume that the treatment increases *BPday* by $1, we're very likely to end up with a CI that includes zero and then conclude that there's no effect. At the leftward end of the curve, our simulated significance is equal to zero, instead of the expected 5%. Let's discuss what causes that and whether we should worry about it.

### Understanding the power/significance trade-off

As I mentioned in the last chapter, if our data is "well-behaved" (i.e., normally distributed, allocated between experimental groups purely at random, etc.) and there is no true effect, we would expect a 90%-CI to include zero 90% of the time, be strictly above it 5% of the time, and be strictly under it 5% of the time. Here, because of the stratified randomization, our false positive rate appears lower than 5%: I observed none whatsoever in 500 simulations. By reducing the noise in our data, stratified randomization also reduces the risk of false positives.

That's nice but in this case it may be too much of a good thing, because it also brings down the power curve for small positive effects up to 1, as we can see in Figure 9-3. Let's put it in a different way: if we had a 5% chance of assuming there is an effect when there is none, then we would have at least a 5% chance of assuming that there is an effect when there is a small one. In that sense, significance gives us some "free" power for low effect sizes.

Let's compare the previous power curve with the power curves for lower confidence levels. By definition, this will give us narrower confidence intervals, meaning a higher significance and a higher power, especially for small effect sizes (Figure 9-4).

As you can see, with a 40%-CI, we only get a small increase in significance but a large increase in power, with a power of approximately 50% to detect an effect size of 0.5.

Does that mean that we should use a 40%-CI instead of a 90%-CI? It depends. Let's get back to the business problem. Our business partners have asked for a power of 90% for an effect size of 2 because they're not interested in going through the hassle of implementing either of the treatments if the benefits are lower than that. Thus, capturing a true effect size of 0.5 with a CI that doesn't include zero, or having a CI that includes zero, is essentially the same thing from a business perspective. Either way, no treatment will be implemented. Therefore, the power curve for the 90%-CI better reflects our business goals.

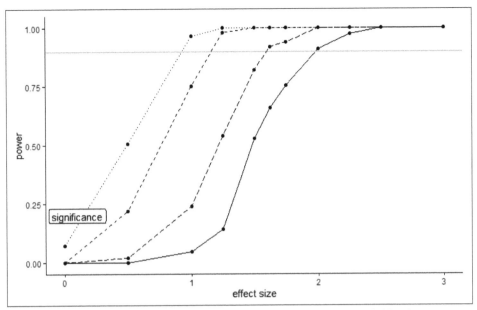

*Figure 9-4. Comparison of power curves for confidence levels 0.90 (solid line), 0.80 (long-dash line), 0.60 (dashed line), and 0.40 (dotted line)*

On the other hand, with an experiment like the "1-click button" in Chapter 8, costs and risks of implementation are limited. The 90% power threshold is just a baseline, and virtually any strictly positive effect would warrant implementation. In such a situation, increasing power for small effect sizes is probably worth a slight increase in significance.

More broadly, when your data or your experimental design diverge from the standard framework, power analysis stops being a simple matter of plugging conventional numbers into a formula and requires understanding what's going on and making judgment calls about the right decisions. Fortunately, power curves offer a great tool to visualize the possible outcomes of an experiment under different scenarios and different decision rules.

# Analyzing and Interpreting Experimental Results

Once we've run the experiment, we can analyze its results. Our target metric, average booking profit per day, is continuous and not binary; therefore, the two appropriate methods are the T-test of means and linear regression. I will refer you to Gerber and Green (2012) if you want to learn more about the T-test, and I'll cover linear regression.

Before getting into the quantitative analysis, remember that we could not force owners to have a two-night minimum duration or to agree to a reduction in the duration of the cleaning window in exchange for free cleaning services. We could only offer them the opportunity to opt in, which some took and others did not. In technical terms, this approach is called an *encouragement design*, because we're encouraging subjects to take up our offer.

Encouragement designs are very common, but they introduce some additional considerations, because we now have two categories of people in a treatment group: those who opted in and those who didn't. For practical purposes, this means we have two different questions we can try to answer:

- What would happen if we offered the possibility of opting into the treatment to our entire owner population?
- What would happen if we enforced the treatment on our entire owner population, without giving them the option of opting out?

The answer to the first question is called the *intention-to-treat* (ITT) estimate because we intend for people to be treated but we don't make it mandatory. The second question is more complex, and we can't answer it fully based only on an encouragement design (or at least it requires additional assumptions), but we can get a closer approximation than the ITT estimate, with the *complier average causal effect* (CACE) estimate.

Let's calculate both of these estimates in turn.

## Intention-to-Treat Estimate for Encouragement Intervention

Let's first calculate the ITT estimate, which will be very easy: it's simply the coefficient for the effect of the experimental assignment, as we've calculated in the previous chapter. Should we factor in the fact that the majority of the owners in the treatment groups did not opt in? No. The fact that the ITT coefficient is diluted by people who opted out is a feature, not a bug: the same dilution would happen at a larger scale.

Let's subtract $10 for the owners in the cleaning group who opted in, to account for the extra cost, then run a linear regression. We could apply our metric functions separately, but I prefer running the whole regression at once to be able to see the other coefficients as well:

```Python
## Python (output not shown)
exp_data_reg_df = exp_data_df.copy()
exp_data_reg_df.BPday = np.where((exp_data_reg_df.compliant == 1) & \
                                 (exp_data_reg_df.grp == 'treat2'),
                                 exp_data_reg_df.BPday -10,
                                 exp_data_reg_df.BPday)
```

```
print(ols("BPday~sq_ft+tier+avg_review+grp",
          data=exp_data_reg_df).fit(disp=0).summary())

## R
> exp_data_reg <- exp_data %>%
    mutate(BPday = BPday - ifelse(grp=="treat2" & compliant, 10,0))
> lin_model <- lm(BPday~sq_ft+tier+avg_review+grp, data = exp_data_reg)
> summary(lin_model)

...
Coefficients:
            Estimate Std. Error t value       Pr(>|t|)
(Intercept) 19.232831   3.573522   5.382 0.0000000854103 ***
sq_ft        0.006846   0.003726   1.838         0.0663 .
tier2        1.059599   0.840598   1.261         0.2077
tier1        5.170473   1.036066   4.990 0.0000006728868 ***
avg_review   1.692557   0.253566   6.675 0.0000000000347 ***
grptreat1    0.966938   0.888683   1.088         0.2767
grptreat2   -0.172594   0.888391  -0.194         0.8460
...
```

We do a regression of our variable of interest, booked profit per day, on the square footage of the property, the city tier, the average of the customer reviews, and the experimental groups. The coefficient for *Grptreat1* refers to the minimum-duration treatment while *Grptreat2* refers to the free-cleaning treatment.

The first treatment increases *BPday* by approximately $0.97 on average, but the p-value is moderately high, at approximately 0.27. This suggests that the coefficient may not be significantly different from zero, and indeed the corresponding Bootstrap 90%-CI is approximately [0.002; 2.66].

 If you were to run a T-test comparing the first treatment group with the control group, you'd find that the absolute value of the test statistic is 0.96, close to the coefficient in the regression we just ran. Similarly, the raw difference in *BPday* averages between the control group and the first treatment group is approximately 0.85. Should that surprise us? No. Thanks to stratification, our experimental groups are very well balanced and therefore other independent variables have the same average effect across groups. This means that even metrics that don't account for covariates are unbiased (their p-values would be off, however, because they don't take into account the stratification).

The second treatment decreases *BPday* by approximately $0.17, after cost, not a great value proposition. The corresponding CI is [-2.23; 1.61].

Remember that our business partners are interested only in implementing an intervention if it generates $2 of additional BP/day above costs. On the face of it, this would preclude implementing the minimum-duration intervention—not just because the statistical significance is borderline, but primarily because of the lack of economic significance. Even if the lower bound of the confidence interval was squarely above zero, this would still not change our business partners' decision.

If this had to be the end of the story, our business partners would not implement either of the encouragement interventions. However, it might be worth considering what would happen if we enforced the minimum-duration treatment across the board without offering people the option to opt out.

## Complier Average Causal Estimate for Mandatory Intervention

When we have an encouragement design, can we estimate the effect of making the treatment mandatory? A tempting but incorrect way to answer it would be to compare the value of the business metric for the opt-in category (a.k.a. the "treated") on the one hand, with the value for the opt-out category and the control group on the other hand, lumping the latter two together as "untreated." One could assume that this comparison would reflect the expected outcome of enforcing the treatment across the board, i.e., imposing a two-night minimum in hot markets or unilaterally shortening the cleaning window while offering free cleaning, regardless of owners' preferences. It does not, because opting in to the treatment is not randomized and is likely to be confounded. Within the treatment groups, it is plausible that the owners who opt in are in some regards different from the owners who don't, e.g., they have financial needs or other characteristics that make them give more attention and effort to their property (Figure 9-5).

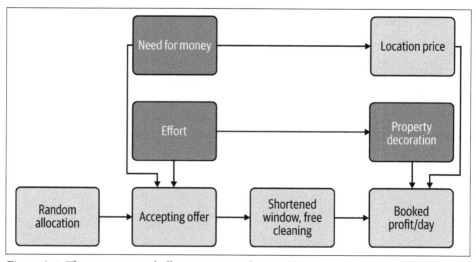

*Figure 9-5. The experimental allocation is randomized but accepting the free-cleaning treatment is not*

If that CD is correct, then accepting the free-cleaning offer is correlated with behaviors that increase booked profit per day and bias upward our coefficient. In other words, we would wrongly attribute to the offer some of the effect of those behaviors if we compared people who opt in with people who don't. The random allocation ensures that comparisons between experimental groups are unbiased, but it doesn't guarantee anything for subgroups further down the line.

 In the case of email A/B tests, this limitation on the effect of randomization means that rates (e.g., opening rate, click-through rate, etc.) should all have as numerator the number of people in the experimental group and not the number of people from the previous stage. If 50% of people open your email and 50% of those who open the email click through, the click-through rate should be expressed as 25%, not 50%.

In an encouragement design, we intend for the people in the treatment group to opt in and get treated, but we also intend for people in the control group to *not* be treated. However, in certain situations we can't prevent them from accessing the treatment. In our present example, the free-cleaning treatment has features that are entirely under our control: property owners may use professional cleaning services but they have to pay for it, and the window between bookings is baked in the software. Therefore no one outside of that treatment group can access that precise treatment. Things are less clear-cut however with the two-night minimum treatment: property owners outside of that treatment group may unofficially enforce a two-night minimum by rejecting single-night requested bookings (Figure 9-6).

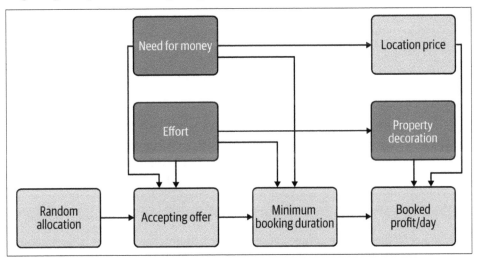

*Figure 9-6. The experimental allocation is randomized but accepting the minimum booking treatment is not, and it can happen outside of the treatment group*

With the minimum booking treatment, there are four possible cases for owners that we can observe:

A. Being in the control group and not having a two-night minimum

B. Being in the control group and having a two-night minimum nonetheless

C. Being in the treatment group and having a two-night minimum

D. Being in the treatment group and not having a two-night minimum nonetheless

This categorization doesn't yet answer our question but it's giving us some important building blocks to distinguish between the effect of the treatment per se and the effect of unobserved factors conducive to setting up a two-night minimum. Let's imagine for a second that we could observe these factors, and rank all the owners in decreasing order of these factors. Because of randomization, we can assume that the control and the treatment groups are reasonably identical in terms of the distribution of unobserved factors (Figure 9-7).

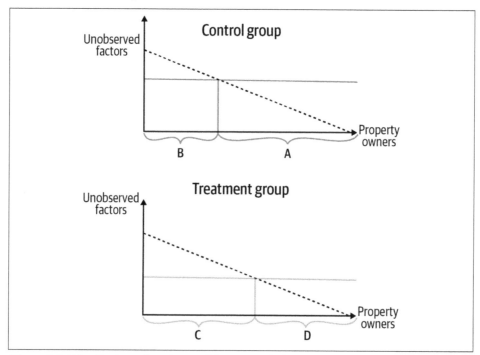

Figure 9-7. Distribution of unobserved factors and observed behaviors in the two groups

All owners in the control group whose value for unobserved factors is high enough will implement a two-night minimum (group B) and all other owners in the control group won't (group A). For the treatment group, we can reasonably assume that the owners with a high factor value are still implementing a two-night minimum and that our encouragement intervention simply lowers the threshold, by getting some owners who wouldn't otherwise to implement it (together they form the group C). Finally, owners who are too low still don't implement it, despite our best efforts (group D).

In econometrics lingo, subjects who would always be treated regardless of their experimental assignment (group B in the control group and the corresponding part of group C in the treatment group) are called *always-takers*. Subjects who would never be treated regardless of their experimental assignment (group D in the treatment group and the corresponding part of group A in the control group) are predictably called *never-takers*. Subjects who are treated if and only if they are in the treatment group (the overlap between group A and group C)" are called *compliers*.

Theoretically, you could have a fourth category, namely subjects who are treated if and only if they are in the control group. They are called *defiers* because they are always doing the exact opposite of what we want them to do. The technical term for this in psychology is *reactance*. While it can happen in real life (cough, teenagers, cough), it's rarely a concern in business settings, unless you're trying to force people to do things they don't want to do and I won't help you with that.

By definition, we can't observe the unobserved factors in our experiment, which means that we can only identify with certainty two groups: the always-takers assigned to control (B) and the never-takers assigned to treatment (D). We can't know if the owners in the treatment group implementing the two-night minimum (C) are always-takers or compliers, and we can't know if the owners in the control group not implementing it (A) are compliers or never-takers. However, and this is where the trick is, we can net out the always-takers and never-takers across experimental groups to measure the effect of the treatment on the compliers, which is called the *complier average causal effect* (CACE). The formula for the CACE is very simple:[3]

$$CACE = \frac{ITT}{P(treated|TG) - P(treated|CG)}$$

In other words, to determine the effect of the treatment on the compliers, we simply need to weight our previous ITT estimate by a measure of the noncompliance in our experiment: if we have full compliance in both groups, meaning that no one gets access to the treatment in the control group (P(treated|CG) = 0) and everyone is treated in the treatment group (P(treated|TG) = 1), this simplifies to the ITT estimate. Very often with encouragement designs, we can prevent people in the control group from accessing the treatment, but only a fraction of the people in the treatment group are actually treated. In that case, the CACE is a multiple of the ITT: if only 10% of the

---

3 If you're curious where it comes from, the derivation is available in the book's GitHub repo (*https://github.com/FlorentBuissonOReilly/BehavioralDataAnalysis*).

people in the treatment group are treated, then our effect is very diluted and our CACE is equal to 10 times the ITT.

The CACE is very useful in two regards: first, it gives us an estimate of the effect of implementing the treatment across the board, without the possibility of opting out. Second, looking at the relationship between the ITT and the CACE allows us to distinguish between two possible situations:

- The ITT is low, but $P(treated|TG)$ - $P(treated|CG)$ is high, meaning that the intervention has a low impact on compliers but a high level of compliance.
- Conversely, the ITT is high, but $P(treated|TG)$ - $P(treated|CG)$ is low, meaning that the intervention has a high impact on compliers but a low level of compliance.

In the first case, we would focus our effort on increasing the effectiveness of the intervention whereas in the second case we would focus on increasing the take-up rate, possibly by making the intervention mandatory. These insights can also help us explore alternative designs: maybe 8 hours is too short, but 12 hours would be acceptable? Maybe we don't need to offer *free* cleaning, simply suggesting a reputable service provider to owners would be enough?

In our present experiment, the uptake rates are quite low in the treatment groups, at about 20% on average:

```
## R (output not shown)
> exp_data_reg %>%
    group_by(grp) %>%
    summarise(compliance_rate = mean(compliant))

## Python
exp_data_reg_df.groupby('grp').agg(compliance_rate = ('compliant', 'mean'))
Out[15]:
        compliance_rate
grp
ctrl              1.000
treat1            0.238
treat2            0.166
```

This means that our CACE estimate for the effect of the minimum-duration treatment will be significantly higher than the ITT estimate:

$$CACE_1 = ITT_1/ComplianceRate_1 = 0.97/0.24 \approx 4.06$$

Now, that's a much more interesting value. The low uptake rate and high CACE suggests that our intervention is fundamentally sound and does generate value when implemented. We can either try to increase the uptake rate by changing the design, or make the intervention mandatory.

---

The CACE has a very neat but narrow interpretation: because we're (implicitly) comparing the same people, the compliers, across the control and the treatment group, our estimate of the effect of the treatment is unbiased. We're not inadvertently capturing the influence of other factors. However, we're measuring it only for that narrow slice of our population, so generalizability is not a given. Compliers might have characteristics that *interact* with our treatment. That is, they might have traits that influence not (or not only) whether they take up the treatment, but how much the treatment affects them. This is where we need to go from a causal to a behavioral lens: is our treatment a tide that lifts all boats, or do the people involved matter? For example, we'll see in the next chapter the example of a talk path in call centers. In that case, complying doesn't just mean applying the treatment, but exerting effort to do so convincingly and not just going through the motions.

Here our intervention is implemented through AirCnC's website. A two-night minimum is a two-night minimum, whoever you're renting from. This means we can be confident that our treatment would be implemented as planned if rolled out across the board, and we can give our business partners the green light to do so.

# Conclusion

In the previous chapter, we had to randomize our experimental assignment "on the fly" as customers connected to the website. In this chapter, we were able to do the random assignment all at once, and therefore *stratify* (a.k.a. *block*) our sample by creating pairs of similar subjects, one of which is assigned to the control group and the other to the treatment group. While this added a layer of complexity, it also significantly increased the effectiveness (in statistical terms, the power) of our experiment. Once you get familiar with stratification, you'll appreciate the ability to extract insights even from small samples.

We also introduced a second complication: our experimental intervention was an *encouragement* treatment. We offered possibilities to owners, but we couldn't force them to take them up, and the uptake was not random. In situations like this, we can easily measure the effect of the encouragement intervention per se, but measuring the effect of accepting the offer (a.k.a. opting in to the treatment) is trickier. Fortunately, we have in the CACE an unbiased estimate of that effect for the compliers in our experimental population. When we can assume the absence of interaction between personal characteristics and the treatment, the CACE can be generalized to our entire experimental population. Even when we can't generalize that far, it provides a less biased estimate than simply comparing the control and treatment group (i.e., the intention-to-treat estimate).

Finally, we had multiple treatments. This did not change anything fundamentally, but it also added some complexity. I would recommend starting your experimentation journey with only one treatment, but I believe in the longer run you'll come to appreciate the organizational "fixed costs" of running an experiment: getting approval from all stakeholders (business partners, legal department, etc.) and getting the technology and data pipeline in place takes barely more time with two treatments than with one. Therefore, running experiments with multiple treatments at once is a key step in increasing the number of treatments tested in a year.

# CHAPTER 10

# Cluster Randomization and Hierarchical Modeling

Our last experiment, while conceptually simple, will illustrate some of the logistical and statistical difficulties of experimenting in business. AirCnC has 10 customer call centers spread across the country, where representatives handle any issue that might come up in the course of a booking (e.g., the payment did not go through, the property doesn't look like the pictures, etc.). Having read an article in the *Harvard Business Review* (HBR) about customer service,[1] the VP of customer service has decided to implement a change in standard operating procedures (SOP): instead of apologizing repeatedly when something went wrong, the call center reps should apologize at the beginning of the interaction, then get into "problem-solving mode," then end up offering several options to the customer.

This experiment presents multiple challenges: due to logistical constraints, we'll be able to randomize treatment only at the level of call centers and not reps, and we'll have difficulties enforcing and measuring compliance. This certainly doesn't mean that we can't or shouldn't run an experiment!

Regarding the randomization constraint, we'll see that this makes the standard linear regression algorithm inappropriate and that we should use hierarchical linear modeling (HLM) instead.

---

1 "'Sorry' Is Not Enough," *Harvard Business Review*, Jan.–Feb. 2018.

As before, our approach will be:

- Planning the experiment
- Determining random assignment and sample size/power
- Analyzing the experiment

# Planning the Experiment

In this section, I'll go briskly through our theory of change to provide you with some necessary context and behavioral grounding:

1. First, the business goal and target metric
2. Next, the definition of our intervention
3. Finally, the behavioral logic that connects them

## Business Goal and Target Metric

Based on the HBR article, our criterion for success or target metric appears straightforward: customer satisfaction as measured by a one-question survey administered by email after the phone call. However, we'll see in a minute that there are complications, so we'll need to revisit it after discussing what we're testing.

## Definition of the Intervention

The treatment we're testing will be whether the reps have been trained in the new SOP and instructed to implement it.

The first difficulty is in the implementation of the treatment. We know from past experience that asking the reps to apply different SOPs to different customers is very challenging: asking reps to switch processes at random between calls increases their cognitive load and the risk of noncompliance. Therefore, we'll have to train some reps and instruct them to use the new SOP for all of their calls, while keeping other reps on the old SOP.

Even with that correction, compliance remains at risk: reps in the treatment group may implement the new SOP inconsistently or even not at all, while reps in the control group may also apply the old SOP inconsistently. Obviously, this would muddle our analysis and make the treatment appear less different from the control group than it is. One way to mitigate this issue is to first observe current compliance with the SOP in place by listening to calls, then run a pilot study, where we select a few reps, train them, and observe compliance with the new SOP. Debriefing the reps in the pilot study after the fact can help identify misunderstandings and obstacles to compliance. Unfortunately, it is generally impossible to have 100% compliance in an experiment where human beings are delivering or choosing the treatment. The best we can do is to try to measure compliance and take it into account when drawing conclusions.

Finally, there is a risk of "leakage" between our control and our treatment groups. Reps are human beings, and reps in a given call center interact and chat. Given that reps are incentivized on the average monthly customer satisfaction (CSAT) for their calls, if reps in the treatment group started seeing significantly better results, there is a risk that reps in the control group of the same call center would start changing their procedure. Having some people in the control group apply the treatment would muddle the comparison for the two groups and make the difference appear smaller than it really is. Therefore, we'll apply the treatment at the call center level: all reps in a given call center will either be in the treatment group or in the control group.

Applying the treatment at the call center level instead of at the call level has implications for our criterion for success. If our unit of randomization is the call center, should we measure the CSAT at the call center level? This would seem logical, but it would mean that we can't use any information about individual reps or individual calls. On the other hand, measuring average CSAT at the rep level or even CSAT at the call level would allow us to use more information, but it is problematic for two reasons:

- First, if we were to disregard the fact that randomization was not done at the call level and use standard power analysis, our results would be biased because randomization is unavoidably correlated with the call center variable; adding more calls in our sample would not change the fact that we have only 10 call centers and therefore only 10 randomization units.

- Second, in our data analysis, we would run into trouble due to the nested nature of the data: assuming that each rep belongs to one and only one call center, there will be multicollinearity between our call center variable and our rep variable (e.g., we can add 1 to the coefficient for the first call center and subtract 1 from the coefficients for all the reps in that call center without changing the results of the regression; therefore the coefficients for the regression are essentially undetermined).

Fortunately, there is a simple solution to this problem: we'll use a hierarchical model, which recognizes the nested structure of our data and handles it appropriately, while allowing us to use explanatory variables down to the call level.[2] For our purposes, we won't get into statistical details and we'll only see how to run the corresponding code

---

2 If you want to learn more about this type of models, Gelman and Hill (2006) is the classic reference on the topic.

and interpret the results. A hierarchical model is a general framework that can be applied to linear and logistic regression, so we'll still be in known territory.

## Behavioral Logic

Finally, the logic for success for this experiment is simple: the new SOP will make customers feel better during the interaction, which will translate into a higher measured CSAT (Figure 10-1).

*Figure 10-1. Causal logic for our experiment*

# Data and Packages

The GitHub folder for this chapter (*https://oreil.ly/BehavioralDataAnalysisCh10*) contains two CSV files with the variables listed in Table 10-1. The check mark (✓) indicates the variables present in that file, while the cross (✗) indicates the variables that are not present.

*Table 10-1. Variables in our data*

|  | Variable description | chap10-historical_data.csv | chap10-experimental_data.csv |
|---|---|---|---|
| *Center_ID* | Categorical variable for the 10 call centers | ✓ | ✓ |
| *Rep_ID* | Categorical variable for the 193 call center reps | ✓ | ✓ |
| *Age* | Age of customer calling, 20-60 | ✓ | ✓ |
| *Reason* | Reason for call, "payment"/"property" | ✓ | ✓ |
| *Call_CSAT* | Customer satisfaction with call, 0-10 | ✓ | ✓ |
| *Grp* | Experimental assignment, "ctrl"/"treat" | ✗ | ✓ |

Note that the two data sets also contain the binary variable *M6Spend*, the amount spent on subsequent bookings within six months of a given booking. This variable will be used in Chapter 11 only.

In this chapter, we'll use the following packages in addition to the common ones:

```R
## R
library(lme4) # For hierarchical modeling
library(lmerTest) # For additional diagnostics of hierarchical modeling
install_github("d4ndo/binaryLogic")
library(binaryLogic) # For function as.binary()

## Python
# To rescale numeric variables
```

```
from sklearn.preprocessing import MinMaxScaler
# To one-hot encode cat. variables
from sklearn.preprocessing import OneHotEncoder"
```

# Introduction to Hierarchical Modeling

Hierarchical models (HMs) can be used when you have categorical variables in your data:

- Customer transactions across multiple stores
- Rental properties across multiple states
- Etc.

Some situations call for HMs because you can't use traditional categorical variables. The main one is if you have a categorical variable that is dependent on another categorical variable (e.g., Vegetarian = {"yes," "no"} and Flavor = {"ham," "turkey," "tofu," "cheese"}), a.k.a. "nested" categorical variables. Then, multicollinearity issues make HMs the way to go. That's also why they are called "hierarchical" models, even though they can be applied to non-nested categories as well.

Beyond that, HMs also offer a more robust alternative if you have a categorical variable with a large number of categories, such as the call center rep ID in our example, and especially if some of the categories have very few rows in your data. Without getting into too much detail, this robustness comes from the way coefficients in HMs incorporate some information from other rows, which brings them closer to the overall average. Let's imagine that we had in our data a call center rep having answered only one call, with an exceptionally bad CSAT that is clearly an outlier. With only one call for that rep, we don't know whether the rep or the call is the outlier. A categorical variable would assign 100% of the "outlier-ness" to the rep, whereas an HM would split it between the rep and the call, i.e., we would expect the rep to have lower-than-average CSAT with other calls, but not as extreme as the observed call.

Finally, in situations where both categorical variables and HMs could be applied (which is basically any situation where you have a categorical variable with a few, non-nested, categories!), there are some nuances in interpretation that may make you prefer one or the other. Conceptually, a categorical variable is a partition of your data into groups with intrinsic differences between them that we want to understand, whereas an HM treats groups as random draws from a potentially infinite distribution of groups. AirCnC has 30 call centers, but it could have been 10 or 50 instead, and we're not interested in the differences between call center number 3 and call center number 28. On the other hand, we'd like to know whether calls for payment reasons have a higher or lower average CSAT than calls related to property issues, and we wouldn't be satisfied with just knowing that the standard deviation between groups is 0.3. But again, these are nuances of interpretation, so don't think too much about it.

## R Code

Let's review the syntax for hierarchical modeling in a simple context, by looking at the determinants of call CSAT in our historical data, leaving the *Rep_ID* variable aside for now. The R code is as follows:

```
## R
> hlm_mod <- lmer(data=hist_data, call_CSAT ~ reason + age + (1|center_ID))
> summary(hlm_mod)
Linear mixed model fit by REML. t-tests use Satterthwaite's method
   ['lmerModLmerTest']
Formula: call_CSAT ~ reason + age + (1 | center_ID)
   Data: hist_data

REML criterion at convergence: 2052855

Scaled residuals:
    Min      1Q  Median      3Q     Max
-4.3238 -0.6627 -0.0272  0.6351  4.3114

Random effects:
 Groups     Name        Variance Std.Dev.
 center_ID (Intercept) 1.406    1.186
 Residual              1.122    1.059
Number of obs: 695205, groups:  center_ID, 10

Fixed effects:
                  Estimate    Std. Error      df       t value  Pr(>|t|)
(Intercept)      3.8990856    0.3749857     9.0938797   10.40 0.00000238 ***
reasonproperty   0.1994487    0.0026669 695193.0006122   74.79    < 2e-16 ***
age              0.0200043    0.0001132 695193.0008798  176.75    < 2e-16 ***
---
Signif. codes:  0 '***' 0.001 '**' 0.01 '*' 0.05 '.' 0.1 ' ' 1

Correlation of Fixed Effects:
           (Intr) rsnprp
reasnprprty 0.000
age        -0.011 -0.236
```

The lmer() function has a similar syntax to the traditional lm() function, with one exception: we need to enter the clustering variable, here center_ID, between parentheses and preceded by 1|. This allows the intercept of our regression to vary from one call center to another. Therefore, we have one coefficient for each call center; you can think of these coefficients as similar to the coefficients we would get in a standard linear regression with a dummy for each call center.[3]

The "Random effects" section of the results refers to the clustering variable(s). The coefficients for each call center ID are not displayed in the summary results (they can be accessed with the command coef(hlm_mod)). Instead, we get measures of the variability of our data within call centers and between call centers, in the form of variance and standard deviation. Here, the standard deviation of our data between call centers is 1.185; in other words, if we were to calculate the mean CSAT for each call center and then calculate the standard deviation of the means, we would get the same value as you can check for yourself:

```
## R
> hist_data %>%
    group_by(center_ID)%>%
    summarize(call_CSAT = mean(call_CSAT)) %>%
    summarize(sd = sd(call_CSAT))
`summarise()` ungrouping output (override with `.groups` argument)
# A tibble: 1 x 1
     sd
  <dbl>
1  1.18
```

The standard deviation of the residuals, here 1.059, indicates how much variability there is left in our data after accounting for the effect of call centers. Comparing the two standard deviations, we can see that the call center effects represent more than half of the variability in our data.

The "Fixed effects" section of the results should look familiar: it indicates the coefficients for the call level variables. Here, we can see that customers calling for a "property" issue have on average a CSAT 0.199 higher than customers calling for a "payment" issue, and that each year of additional age for our customers adds on average 0.020 to the call CSAT.

Let's then include the rep_ID variable as a clustering variable nested under the center_ID variable:

```
## R
> hlm_mod2 <- lmer(data=hist_data,
                  call_CSAT ~ reason + age + (1|center_ID/rep_ID),
                  control = lmerControl(optimizer ="Nelder_Mead"))
> summary(hlm_mod2)
Linear mixed model fit by REML. t-tests use Satterthwaite's method
   ['lmerModLmerTest']
Formula: call_CSAT ~ reason + age + (1 | center_ID/rep_ID)
```

---

3 If you really want to know, these coefficients are calculated as a weighted average of the mean CSAT in a call center and the mean CSAT across our whole data.

```
     Data: hist_data
Control: lmerControl(optimizer = "Nelder_Mead")

REML criterion at convergence: 1320850

Scaled residuals:
    Min      1Q  Median      3Q     Max
-5.0373 -0.6712 -0.0003  0.6708  4.6878

Random effects:
 Groups             Name        Variance Std.Dev.
 rep_ID:center_ID (Intercept) 0.7696   0.8772
 center_ID        (Intercept) 1.3582   1.1654
 Residual                     0.3904   0.6249
Number of obs: 695205, groups:  rep_ID:center_ID, 193; center_ID, 10

Fixed effects:
                 Estimate  Std. Error          df t value  Pr(>|t|)
(Intercept)    3.90099487  0.37397956  8.73974599   10.43 0.00000316 ***
reasonproperty 0.19952547  0.00157368 695010.05594912 126.79  < 2e-16 ***
age            0.01992162  0.00006678 695010.05053170 298.30  < 2e-16 ***
---
Signif. codes:  0 '***' 0.001 '**' 0.01 '*' 0.05 '.' 0.1 ' ' 1

Correlation of Fixed Effects:
            (Intr) rsnprp
reasnprprty  0.000
age         -0.007 -0.236
```

As you can see, this is done by adding `rep_ID` as a clustering variable after `center_ID`, separating them with /. Also note that I was getting a warning that the model had failed to converge, so I changed the optimizer algorithm to `"Nelder_Mead"`.[4] The coefficients for the fixed effects are slightly different, but not that much.

## Python Code

Though it's more concise, Python code works similarly. The main difference is that the groups are expressed with `groups = hist_data_df["center_ID"]`:

```
## Python
mixed = smf.mixedlm("call_CSAT ~ reason + age", data = hist_data_df,
                    groups = hist_data_df["center_ID"])
print(mixed.fit().summary())
            Mixed Linear Model Regression Results
============================================================
Model:               MixedLM Dependent Variable: call_CSAT
No. Observations:    695205  Method:               REML
```

---

4 As always with numerical simulations, your mileage may vary. Thanks to Jessica Jakubowski for suggesting an alternative specification: `lmerControl(optimizer ="bobyqa", optCtrl=list(maxfun=2e5))`.

```
No. Groups:           10      Scale:              1.1217
Min. group size:   54203      Log-Likelihood:     -1026427.7247
Max. group size:   79250      Converged:          Yes
Mean group size:  69520.5
-----------------------------------------------------------------
                     Coef. Std.Err.    z    P>|z| [0.025 0.975]
-----------------------------------------------------------------
Intercept            3.899    0.335 11.641 0.000  3.243  4.556
reason[T.property]   0.199    0.003 74.786 0.000  0.194  0.205
age                  0.020    0.000 176.747 0.000 0.020  0.020
Group Var            1.122    0.407
=================================================================
```

The coefficients for the fixed effects (i.e., the intercept, the reason for the call and age) are identical to the R code. The coefficient for the variance of the random effect is expressed at the bottom of the fixed effects. At 1.122, it's slightly different from the R value, due to differences in algorithms, but it won't affect the coefficients we care about.

Using nested clustering variables also has a different syntax in Python. We need to express the lower-level, nested, variable in a separate formula (the "variance components formula," which I abbreviated as vcf):

```
## Python
vcf = {"rep_ID": "0+C(rep_ID)"}
mixed2 = smf.mixedlm("call_CSAT ~ reason + age",
                data = hist_data_df,
                groups = hist_data_df["center_ID"],
                vc_formula=vcf)
print(mixed2.fit().summary())
            Mixed Linear Model Regression Results
============================================================
Model:             MixedLM Dependent Variable:  call_CSAT
No. Observations:  695205  Method:               REML
No. Groups:        10      Scale:                0.3904
Min. group size:   54203   Log-Likelihood:       -660498.6462
Max. group size:   79250   Converged:            Yes
Mean group size:  69520.5
------------------------------------------------------------
                     Coef. Std.Err.    z    P>|z| [0.025 0.975]
------------------------------------------------------------
Intercept            3.874    0.099 38.992 0.000  3.679  4.069
reason[T.property]   0.200    0.002 126.789 0.000 0.196  0.203
age                  0.020    0.000 298.301 0.000 0.020  0.020
rep_ID Var           1.904    0.303
============================================================
```

The syntax for the variance components formula is a bit esoteric but the intuition is straightforward. The formula itself is a dictionary with each of the nested variables as key. The value attached to each key indicates whether we want that variable to have a random intercept or a random slope (random here means "varying by category"). A random intercept is the HM equivalent of a *categorical* variable and is expressed as `"0+C(var)"`, where `var` is the name of the nested variable, i.e., the same as the key. Random slopes are beyond the scope of this book, but for example if you wanted the relationship between age and call satisfaction to have a different slope for each rep, the variance component formula would be `vcf = {"rep_ID": "0+C(rep_ID)", "age":"0+age"}`, without a `C()` in the second case.

# Determining Random Assignment and Sample Size/Power

Now that we have planned the qualitative aspects of our experiment, we need to determine the random assignment we'll use as well as our sample size and power. In our two previous experiments (Chapter 8 and Chapter 9), we had some target effect size and statistical power, and we chose our sample size accordingly. Here, we'll add a wrinkle by assuming that our business partners are willing to run the experiment only for a month,[5] and the minimum detectable effect they're interested in capturing is 0.6 (i.e., they want to make sure that you have sufficient power to capture an effect of that size, but they're willing to take the risk that the effect size will be lower).

Under these constraints, the question becomes: how much power do we have to capture a difference of that amount with that sample? In other words, assuming that the difference is indeed equal to 0.6, what is the probability that our decision rule will conclude that the treatment is indeed better than the control?

As mentioned earlier, we'll be using a hierarchical regression to analyze our data and that will complicate our power analysis a bit, but let's first briefly review the process for random assignment.

## Random Assignment

Even though we don't know ahead of time which customers are going to call, it doesn't matter for the random assignment because we'll do it at the call center level. Therefore, we can do it in advance, assigning control and treatment groups all at once. With clustered experiments like this one, stratification is especially useful because we have so few actual units to randomize. Here, we're randomizing at the level of call centers, so we would want to stratify based on the centers' characteristics, such as number of reps and average values of the call metrics. Let's first aggregate data at the center level, like we did in Chapter 9.

---

[5] Does that suck for your experimental design? Totally. Is that unrealistic? Absolutely not, unfortunately. As we used to say when I was a consultant, the client is always the client.

```
## R

# Aggregating data to center level
center_data <- hist_data %>%
  group_by(center_ID) %>%
    summarise(nreps = n_distinct(rep_ID),
              avg_call_CSAT = mean(call_CSAT),
              avg_age = mean(age),
              pct_reason_pmt = sum(reason == 'payment')/n()) %>%
      ungroup()

## Python

# Aggregating data to center level
center_data_df = hist_data_df.groupby('center_ID').agg(
      nreps = ('rep_ID', lambda x: x.nunique()),
      avg_call_CSAT = ("call_CSAT", "mean"),
      avg_age=("age", "mean"),
      pct_reason_pmt=('reason',
                      lambda x: sum(1 if r=='payment' else 0 for r in x)/len(x))
      )

#Reformatting variables as needed
center_data_df['nreps'] = center_data_df.nreps.astype(float)
```

In the previous chapter, we only care about the final assignments to the different experimental groups, but in this one, we'll stop at an intermediary step, the pairing of subjects (in this case, centers) into strata.

In R, we'll use the pair() function from the BehavioralDataAnalysis package.

```
## R

> center_pairs <- pair(center_data, id = "center_ID")
> center_pairs
     [,1] [,2]
[1,] "2"  "1"
[2,] "6"  "3"
[3,] "7"  "4"
[4,] "10" "5"
[5,] "9"  "8"
```

The pair() function does the heavy lifting under the hood of paired_assign(), so the two functions have a similar syntax. The first argument is the data to stratify; the second argument is the id variable for the units to pair; the third argument, the number of groups to form, is left to its default value of 2 here. Each row in the output matrix forms a pair. That is, the function determined that centers 1 and 2 were "close" to each other and should form a pair, and so on. We just don't assign the centers to the control and the treatment group yet, as we'll discuss below.

In Python, we don't have a ready-made pair function so we'll create one on the fly using the code from last chapter.

*Example 10-1. Pairing of call centers in Python*

```python
## Python

def strat_prep_fun(dat_df):

    #Extracting components of the data
    num_df = dat_df.copy().loc[:,dat_df.dtypes=='float64'] #Numeric vars
    center_ID = [i for i in dat_df.index]

    #Normalizing all numeric variables to [0,1] ❶
    scaler = MinMaxScaler()
    scaler.fit(num_df)
    num_np = scaler.transform(num_df)

    return center_ID, num_np

def pair_fun(dat_df, K = 2):

    match_len = K - 1 # Number of matches we want to find
    match_idx = match_len - 1 # Accounting for 0-indexing

    center_ID, data_np = strat_prep_fun(dat_df)
    N = len(data_np)

    #Calculate distance matrix ❷
    from scipy.spatial import distance_matrix
    d_mat = distance_matrix(data_np, data_np)
    np.fill_diagonal(d_mat,N+1)
    # Set up variables
    available = [i for i in range(N)]
    available_temp = available.copy()
    matches_lst = []
    lim = int(N/match_len)

    closest = np.argpartition(d_mat, kth=match_idx,axis=1) ❸

    for n in available:
        if len(matches_lst) == lim: break
        if n in available_temp:
            for match_lim in range(match_idx,N-1): ❹
                possible_matches = closest[n,:match_lim].tolist()
                matches = list(set(available_temp) & set(possible_matches))
                if len(matches) == match_len:
                    matches.append(n)
                    matches_lst.append(matches)
                    available_temp \
                    = [m for m in available_temp if m not in matches]
                    break
                else:
                    closest[n,:] = np.argpartition(d_mat[n,:], kth=match_lim)
    #Map center indices to their proper IDs
```

```
    matches_id_lst = [[center_ID[k[0]],center_ID[k[1]]] for k in matches_lst]
    return np.array(matches_id_lst)
pair_fun(center_data_df)
```

❶ We rescale all clustering variables to between 0 and 1.

❷ We calculate the distance matrix between all the centers.

❸ We use the `argpartition()` function from NumPy, which returns for each center its closest neighbors.

❹ For each center, we create a pair with its closest neighbor, or if it's already taken we look for the next closest neighbor until we can form a pair.

## Power Analysis

Using a standard statistical formula for power analysis (in this case it would be the formula for the T-test) would be highly misleading because it would not take into account the correlation that exists in the data. Gelman and Hill (2006) provide some specific statistical formulas for hierarchical models, but I don't want to go down the rabbit hole of accumulating increasingly complex and narrow formulas. As usual, we'll be running simulations as our foolproof approach to power analysis.

Let's first define our metric function:

```
## R
hlm_metric_fun <- function(dat){
  #Estimating treatment coefficient with hierarchical regression
  hlm_mod <- lmer(data=dat,
              call_CSAT ~ reason + age + group + (1|center_ID/rep_ID)
              ,control = lmerControl(optimizer ="Nelder_Mead")
              )
  metric <- fixef(hlm_mod)["grouptreat"]
  return(metric)}

## Python
def hlm_metric_fun(dat_df):
    vcf = {"rep_ID": "0+C(rep_ID)"}
    h_mod = smf.mixedlm("call_CSAT ~ reason + age + group",
                  data = dat_df,
                  groups = dat_df["center_ID"],
                  re_formula='1',
                  vc_formula=vcf)
    coeff = h_mod.fit().fe_params.values[2]
    return coeff
```

This function returns the coefficient for the treatment group from our hierarchical model. As we did in the previous chapters, let's now run the simulations for our power analysis, which you should hopefully be familiar with by now. The only additional thing we need to take into account here is that our data is stratified, a.k.a. clustered. This has two implications.

First, we can't just draw calls from historical data at random. In our experiment, we expect reps to have almost exactly the same number of calls each; on the other hand, a truly random draw would generate some significant variation in the number of calls per rep. We expect reps to handle around 1,200 calls a month; having one rep handle 1,000 calls and another handle 1,400 is much more likely with a truly random draw than in reality. Fortunately, from a programming standpoint, this can easily be resolved by grouping our historical data at the call center and rep level before making a random draw:

```
## R
sample_data %<%- filter(dat, month==m) %>%dplyr::group_by(rep_ID) %>%
    slice_sample(n = Nexp) %>% dplyr::ungroup()

## Python
sample_data_df = sample_data_df.groupby('rep_ID')\
            .sample(n=Ncalls_rep, replace=True)\
            .reset_index(drop = True)
```

## Using permutations when randomness is "limited"

The second implication is at the statistical level and is more profound. We're using stratification to pair similar call centers and assign one from each pair to the control group and the other to the treatment group. This is good, because we reduce the risk that some call center characteristics will bias our analysis. But at the same time, this introduces a fixed effect in our simulations: let's say that call centers 1 and 5 are paired together because they're very similar. Then, however many simulations we run, one of them will be in the control group and the other one will be in the treatment group; we've reduced the total number of possible combinations. With a completely free randomization, there are $10!/(5! * 5!) \approx 252$ different assignments of 10 call centers in equally sized experimental groups, which is already not that many.[6] With stratification, there are only $2^5 \approx 32$ different assignments, because there are two possible assignments for each of the five pairs: (control, treatment) and (treatment, control). This means that even if you were to run 32,000 simulations, you would only see 32 different random allocations at the call center level. Moreover, with only three months worth of historical data, we can only generate three completely different (i.e., mutually exclusive) samples per rep, for a total of $32 * 3 = 96$ different simulations.

This doesn't mean that we should not use stratification; on the contrary, stratification is even more crucial the smaller our experimental population gets! This does imply

---

[6] The exclamation mark indicates the mathematical operator factorial. See this Wikipedia page (*https://oreil.ly/I5PTW*) if you want to better understand the underlying math.

however that it is pretty much pointless and potentially misleading to run many more simulations than you have truly different assignments.

To understand why, let's use a metaphor: imagine a student who decides to increase their vocabulary ahead of a test (e.g., the LSAT). They buy a learner's dictionary and plan to read the definition of a random word in it ten times a day until they've done it a thousand times, to learn a thousand words. But here's the catch: their dictionary has only 96 words in it! This means that however many times the student looks up a random word, their vocabulary cannot increase by more than 96 words. There's certainly value in reading a word's definition more than once, to better understand and memorize it, but that's different from reading the definition of more words. This also means that looking at definitions at random is a very inefficient way to proceed. It is much better to simply go through the 96 words in order.

That logic applies in the same way to simulations: we usually draw at random from our historical data to build a simulated experimental data set, and we (correctly) treat as negligible the probability of several simulations being identical. In the present case, if we had a hundred call centers, each with a thousand reps and ten years of data, we could confidently simulate hundreds, or even thousands, of experiments without worrying. With our limited number of call centers and reps, we're better off going systematically through the limited number of possibilities.

Let's see how to do this in code. We have the call-center pairings (see the R output toward the end of the subsection "Random Assignment" on page 242) and we need to go through the 32 possible permutations of those pairs. The first pair is made up of call centers #2 and #1, so half of the simulations will have #2 in control group and #1 in treatment group, while the other half will have #1 in control group and #2 in the treatment group, and so on. So the first simulation might have as control group the call centers (2, 6, 7, 10, 9) while the second simulation has as control group (1, 6, 7, 10, 9).

We'll use a trick to help us go through the permutations easily. It is not really complex, but it relies on properties of binary numbers that are not intuitive, so brace yourself and bear with me. Any integer can be expressed in binary base as a sequence of zeros and ones. 0 is 0, 1 is 1, 2 is 10, 3 is 11, and so on. These can be left-padded with zeros to have a constant number of digits. We want the number of digits to be equal to the number of pairs, here 5. This means that 0 is 00000, 1 is 00001, 2 is 00010, and 3 is 00011. The largest integer we can express with 5 digits is 31. Note that, and this is not a coincidence, including 0 as 00000, we can express 32 different integers with 5 binary digits, and that 32 is the number of permutations we want to implement. Therefore, we can decide that the first simulation, which we'll call "simulation 00000," has as control group (2, 6, 7, 10, 9). From there, we'll swap a pair between the control and the treatment groups whenever the digit corresponding to the pair in the binary form of the simulation number is a 1. So for example, for simulation 10000, we would swap call centers #2 and #1, giving us the control group (1, 6, 7, 10, 9). Here's where the magic happens: by going from 00000 to 11111, we'll see all the possible permutations of the 5 pairs!

### Code for permutations

Because of the differences in indexing between Python and R (the former starting at 0 and the latter at 1), the code is a bit simpler in Python, so let's start with the corresponding snippet of code:

```
## Python
for perm in range(Nperm):
    bin_str = f'{perm:0{Npairs}b}'   ❶
    idx = np.array([[i for i in range(Npairs)],   ❷
                    [int(d) for d in bin_str]]).T
    treat = [stratified_pairs[tuple(idx[i])] for i in range(Npairs)]   ❸

    sim_data_df = sample_data_df.copy()
    sim_data_df['group'] = 'ctrl'   ❹
    sim_data_df.loc[(sim_data_df.center_ID.isin(treat)),'group']\
        = 'treat'
```

❶  We convert the permutation counter `perm` to a binary string. In Python, there are several ways to do it. I did it here with an F-string. The syntax of an F-string is `f'{exp}'`, where the expression `exp` is evaluated before getting formatted as a string. Within the expression, `Npairs` is also between curly braces, so it's evaluated first before being passed to the expression; after that first evaluation, `exp` is equal to `perm:05b`. The first term on the left of the colon is the number to format; the letter after the colon indicates the format to use, here b for binary; the number immediately to the left of the letter indicates the total number of digits to use (here 5); and finally, any character to the left of that number is to be used for padding (here 0).

❷  We match the digits of the binary string with a counter for the pairs within the `idx` matrix. So "00000" becomes $\begin{pmatrix} 0 & 0 \\ 1 & 0 \\ 2 & 0 \\ 3 & 0 \\ 4 & 0 \end{pmatrix}$ in Python after transposing.

❸  We pass the rows of `idx` as indices to indicate which element of each pair goes into the treatment group. That is, to indicate that the first element of the first pair should go into the treatment group, we pass [0, 0]. With 00000, we always put the first element of each pair in the treatment group. With the last permutation, 11111, we put the second element of each pair in the treatment group, mirroring the allocation for 00000. Taking a more complicated example, for permutation number 7, whose binary format is 00111, we would put in the control group the first element for the first two pairs, and the second element for the last three pairs.

❹ Finally, we update our simulated experimental data set, assigning each row to either the control or treatment group based on its center ID.

The process is identical in R with a few differences in syntax:

```
## R
permutation_gen_fun <- function(i, stratified_pairs){
  Npairs <- nrow(stratified_pairs)
  bin_str <- as.binary(i, n=Npairs) ❶
  idx <- matrix(c(1:Npairs, bin_str), nrow = Npairs)
  idx[,2] <- idx[,2] + 1 ❷
  treat <- stratified_pairs[idx] ❸
  return(treat)}
```

❶ Converting perm to a binary format is done in R with the as.binary() function, which takes as first argument the number to convert and as second argument the total number of digits we want (i.e., the number of pairs, here 5).

❷ Because the indexing starts with 1 and not 0 in R, we need to add 1 to all the elements of the second column in the idx matrix. Thus, for the first permutation, 00000, where the first element of each pair goes into the control group, the idx

matrix is $\begin{pmatrix} 1 & 1 \\ 2 & 1 \\ 3 & 1 \\ 4 & 1 \\ 5 & 1 \end{pmatrix}$. For permutation 11111, the second column would be made up of

2s, and for 00111 it would be 11222.

❸ We pass the rows of idx as indices to indicate which element of each pair goes into the treatment group.

The permutation_gen_fun() function returns a list of the center IDs for the treatment group, which can then be used in the random assignment function.

## Power curve

Now that we have a solution to the problem of limited possible samples, we can get back to our power analysis. Remember that business partners want to run the experiment for no longer than a month, meaning a sample size of about 230,000 calls. Instead of calculating the required sample size for the threshold value of 0.6 points of

CSAT and the desired power, we need to take the sample size as given and calculate what power we have for this threshold value.

Let's first look at statistical significance. Remember that in the previous chapter, our estimator was "underconfident": the 90%-CI included zero more than 90% of the time. Even using a 40%-CI led only to a small number of false positives. Here, we have the opposite problem: our estimator is "overconfident" as the 90%-CI includes zero much less than 90% of the time, and indeed it never includes it: our coverage is null. Figure 10-2 shows the 96 confidence intervals ranked from lowest to highest.

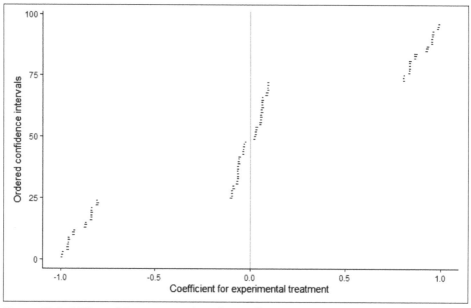

*Figure 10-2. 90% confidence intervals with no effect*

The situation we can see in Figure 10-3 is similar to what we saw in Chapter 7, where having very limited data led to discontinuities in our graphs. Here, the random errors never quite line up in the way that would result in a CI including zero. Instead, we have four tight clusters of CIs, even though the distribution of our CIs is symmetric around zero (i.e., our estimator is unbiased) and half of them are very close to it. From a practical perspective, that means that if we run our experiment, we shouldn't expect the true value to be included in our CI.

This doesn't mean that our experiment is doomed, but that we shouldn't trust our CI bounds and we should rely on our decision rule instead. With the default decision rule of accepting any CI that is strictly positive, our significance is 50%: because half of our CIs are below zero and half above, in half of the cases we would observe a negative coefficient and rightly conclude that the treatment group is no better than the control group. Figure 10-3 plots the power curve with this decision rule for different effect sizes.

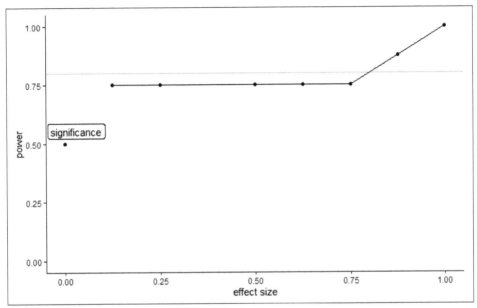

*Figure 10-3. Power curve with decision threshold of 0 for different effect sizes*

As you can see, our power reaches 75% very quickly, basically as soon as the cluster of CIs that was just below zero gets shifted just above it. After that, our power remains constant for a range of values including our threshold effect size of 0.6, until the cluster of strongly negative CIs gets shifted above zero in turn. Then our power is close to 100% for effect sizes of 1 or above. That is, if the true effect is 1 or higher, we're extremely unlikely to see a negative CI.

We could go back to our business partners and tell them that our CIs are unreliable and therefore our risk of false positives is large, but our risk of false negatives is very low. In the present case, we can do better by setting a more stringent decision rule and implementing the intervention only if we observe an effect size of 0.25 or above. Figure 10-4 shows the power curve for that decision rule.

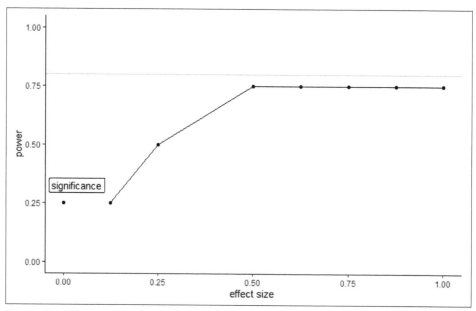

*Figure 10-4. Power curve for different effect sizes with decision threshold of 0.25*

As we can see in Figure 10-4, by increasing our decision threshold, we've lowered the left side of our power curve. This implies a lower significance (i.e., lower risk of false positives) at the cost of a lower power (i.e., higher risk of false negatives) for small effect sizes. However, the right side of our power curve remains mostly unchanged, meaning that our power to detect an effect of 0.6 remains at 75%.

Let's recap what our power analysis told us. Because we plan to use a stratified random assignment with a limited number of effective experimental units (i.e., call centers), our experiment has a rigid structure that constrains possible outcomes. This makes our CIs unreliable by themselves. However, we can adjust our decision rule to a higher threshold (i.e., we'll implement our intervention only if we observe an effect of 0.25 or higher). By doing so, we can reduce the risk of false positives for a null effect size while keeping our power for the target effect size high enough. This remains an underpowered experiment but this is the best we can offer as experimentalists, and our business partners will have to decide how they feel about these odds.

 Note the difference between our decision threshold, 0.25, and our target effect, 0.6. By definition, the power at the decision threshold is always 0.5 and we set out to get as much power as possible for an effect size of 0.6.

# Analyzing the Experiment

Once we have run our experiment, we can collect and analyze the data. Having defined our metric function previously, the analysis is now as simple as applying it to our experimental data, then obtaining a Bootstrap 90%-CI of its value for our experimental data:

```
## R (output not shown)
> coeff <- hlm_metric_fun(exp_data)
> print(coeff)
> hlm_ci_values <- boot_ci(exp_data, hlm_metric_fun)
> print(hlm_ci_values)

## Python
coeff = hlm_metric_fun(exp_data_df)
print(coeff)
hlm_CI = boot_CI_fun(exp_data_df, hlm_metric_fun)
print(hlm_CI)
0.477903237163797
[0.47434045128179986, 0.4815858577196438]
```

Our confidence interval is very narrow and squarely above 0.25. Based on our power analysis, the true effect size is unlikely to be actually within that CI, but it is as likely to be lower as it is to be higher, so our expected effect size is equal to 0.48. Because that is above our decision threshold, we would implement the intervention, even though the expected effect size is less than our target. Interestingly, that confidence interval is much smaller than the one we would obtain based on the normal approximation (i.e., coefficient +/− 1.96 * coefficient standard error), in part because of the stratified randomization.

# Conclusion

This concludes our tour of experimental design. In the last part of the book, we'll see advanced tools that will allow us to dig deeper in the analysis of experimental data, but the call center experiment we've just seen is about as complex as experiments get in real life. Being unable to randomize at the lowest level and having a predetermined amount of time to run an experiment are unpleasant but not infrequent circumstances. Randomizing at the level of an office or a store instead of customers or employees is common, to avoid logistical complications and "leakage" between experimental groups. Leveraging simulations for power analysis and stratification for random assignment becomes pretty much unavoidable if you want to get useful results out of your experiment; hopefully you should now be fully equipped to do so.

Designing and running experiments is in my opinion one of the most fun parts of behavioral science. When everything goes well, you get to measure with clarity the impact of a business initiative or a behavioral science intervention. But getting everything to go well is no small feat in itself. Popular media and business vendors often feed the impression that experimentation can be as simple as "plug and play, check for 5% significance and you're done!" but this is misleading, and I've tried to address several misconceptions that come out of this.

First of all, statistical significance and power are often misunderstood, which can lead to wasted experiments and suboptimal decisions. I believe that eschewing p-values in favor of Bootstrap confidence intervals leads to results and interpretations that are both more correct and more relevant to applied settings.

Second, treating experiments as a pure technology and data analysis problem is easier but less fruitful than adopting a causal-behavioral approach. Using causal diagrams allows you to articulate more clearly what would be a success and what makes you believe your treatment would be successful.

Implementing an experiment in the field is fraught with difficulties (see Bibliography for further resources), and unfortunately each experiment is different, therefore I can only give you some generic advice:

- Running field experiments is an art and science, and nothing can replace experience with a specific context. Start with smaller and simpler experiments at first.

- Start by implementing the treatment on a small pilot group that you then observe for a little while and extensively debrief. This will allow you to ensure as much as possible that people understand the treatment and apply it somewhat correctly and consistently.

- Try to imagine all the ways things could go wrong and to prevent them from happening.

- Recognize that things will go wrong nonetheless, and build flexibility into your experiment (e.g., plan for "buffers" of time, because things will take longer than you think—people might take a week to settle into implementing the treatment correctly, data might come in late, etc.).

# PART V

# Advanced Tools in Behavioral Data Analysis

This is the last part of the book, where everything comes together. We'll see three powerful tools of behavioral data analysis, first moderation in Chapter 11 and then mediation and its offspring instrumental variables (IVs) in Chapter 12.

Moderation is a versatile mathematical tool that allows us to understand interaction effects, as well as effectively and transparently segment our customer population. Mediation allows us to peek into the black box of causal relationships and understand *how* a variable affects another. Finally, I'll use instrumental variables to make good on my promise to measure the impact of customer satisfaction on later customer behaviors.

Moderation, mediation, and IVs are meaty topics and are the subject of lively methodological debates and entire books. But the tools we introduced earlier in the book will make the journey much easier. First, using causal diagrams will give intuitive interpretations for all three of them. Second, using the Bootstrap will allow us to directly build confidence intervals and completely bypass the methodological complications associated with p-values. As a result, we'll be able to draw deep and actionable behavioral insights with one-liners of code.

# CHAPTER 11

# Introduction to Moderation

One of the most gratifying aspects of combining the causal and behavioral perspectives is that things that may seem entirely unrelated under one of them can turn out to be the exact same thing under the other. More simply put, when you have the right hammer, a lot of things *are indeed nails.*

So far, we've used causal diagrams to understand what drives behaviors on average: if temperature increases by one degree, keeping all the relevant other variables constant, by how much do sales of ice cream in C-Mart stands increase? But very often, we're not just interested in that grand average, and we would like to break it down further:

- Does that number apply equally to the stands in Texas and Wisconsin? If not, this means our data shows an opportunity for *segmentation.*

- Does that number apply equally for chocolate and vanilla ice cream? If not, this means that there is an *interaction* between temperature and ice cream flavor.

- Does that number apply equally at low and high temperatures? If not, this means that there is a *nonlinearity* in the effect of temperatures on sales.

The hammer we'll see in this chapter is called moderation analysis by social scientists, and it will allow us to address these three types of questions in the exact same way.

In the first section after reviewing the data and packages for this chapter, we'll do a tour of moderation and see how it can apply to a variety of behavioral situations. Because the math remains the same in all cases, I have gathered all practical and technical considerations to review in the last section.

# Data and Packages

The GitHub folder for this chapter (*https://oreil.ly/BehavioralDataAnalysisCh11*) contains the CSV file *chap11-historical_data.csv* with the variables listed in Table 11-1.

*Table 11-1. Variables in our data*

| Variable name | Variable description |
|---|---|
| Day | Day index, 1-20 |
| Store | Store index, 1-50 |
| Children | Binary 0/1, whether the customer is accompanied by young children |
| Age | Customer age, 20-80 |
| VisitDuration | Duration of store visit in minutes, 3-103 |
| PlayArea | Binary 0/1, store level, whether the store has a play area |
| GroceriesPurchases | Amount spent on groceries purchases during visit in dollars, 0-324 |

In this chapter, we'll only use the common packages so there are no chapter-specific ones.

# Behavioral Varieties of Moderation

The formal definition of moderation is extremely simple: it's the inclusion in a regression of a multiplication between two predictors. For example, I suggested earlier that sales of ice cream may increase more or less per degree of temperature in Texas and Wisconsin; this would be expressed mathematically as follows:

$$IceCreamSales = \beta_t.Temperature + \beta_s.State + \beta_{ts}.(Temperature * State)$$

Moderation can be used to understand all of the following behavioral phenomena, which we'll review in turn:

- Segmentation
- Interactions
- Nonlinearities (a.k.a. self-moderation)

# Segmentation

Building relevant customer segments is a key task in marketing analytics, and more broadly business analytics. We'll review how to do it with observational data, then with experimental data.

## Segmenting observational data

Our starting point will be a C-Mart example: the company has recently introduced play areas in some of its stores and is interested in understanding how it affected customer visit duration. Regression analysis supported by causal diagrams gives us an average causal effect: across all the customer visits in our data, what is the effect on visit duration of having a play area in the store? However, averages can be misleading and hide large differences between segments of our population. For example, it makes sense to assume that the presence of a play area influences visit duration more for customers with children. How should we account for that in our regression? You might think that simply including *Children* as another predictor of *VisitDuration* would do the trick, as in Figure 11-1.

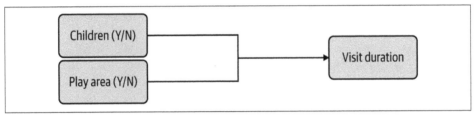

*Figure 11-1. Including Children as a predictor of VisitDuration*

The problem with that approach is that it accounts for the impact of *Children* on *VisitDuration*, *regardless of whether there is a play area or not*, and vice versa: when doing a regression, each variable is examined independently. Each coefficient is set such that overall residual distances are minimized, but the coefficients have to be the same for each variable regardless of the values of other variables. That means that the coefficient, and thus the measured effect, of *PlayArea* on *VisitDuration* is mathematically forced to be a weighted average of its effect for customers with and without children. Likewise, the effect of customers having children is measured as a weighted average of the effects when there is a play area present and when there is not. This can be confirmed by looking at the corresponding equations. The equation for the CD we just drew is shown in Equation 11-1 (including the constant coefficient $\beta_0$, which we usually omit for simplicity).

*Equation 11-1.*

*VisitDuration* = $\beta_0 + \beta_p$.*PlayArea* + $\beta_c$.*Children*

Because *PlayArea* and *Children* are both binary variables, we have four possible cases depending on whether each one of them is equal to 0 or 1:

1. $\beta_0$ is the average visit duration for customers without children in stores without a play area (which we'll abbreviate as C = 0, P = 0 for brevity).

2. $\beta_0 + \beta_c$ is the average visit duration for customers with children in stores without a play area (C = 1, P = 0).

3. $\beta_0 + \beta_p$ is the average visit duration for customers without children in stores with a play area (C = 0, P = 1).

4. $\beta_0 + \beta_p + \beta_c$ is the average visit duration for customers with children in stores with a play area (C = 1, P = 1).

This means that whether a customer has children with them or not doesn't affect the impact of adding a play area, as we can easily check:

If *Children* = 0, the impact of adding a play area is:

*VisitDuration*(C = $_0$, P = 1) - *VisitDuration*(C = 0, P = 0) = $(\beta_0 + \beta_p) - (\beta_0) = \beta_p$

If *Children* = 1, the impact of adding a play area is:

*VisitDuration*(C = 1, P = 1) - *VisitDuration*(C = 1, P = 0) = $(\beta_0 + \beta_c + \beta_p) - (\beta_0 + \beta_c)$
$= \beta_p$

The difference between these two equations (i.e., how much *more* adding a play area increases visit duration for customers with children compared to customers without children) is by definition:

$$[(\beta_0 + \beta_c + \beta_p) - (\beta_0)] - [(\beta_0 + \beta_p) - (\beta_0)] = \beta_p - \beta_p = 0$$

An equivalent way of looking at the problem is that we have four equations, with four corresponding averages, but only three coefficients. Once we have set $\beta_0$, $\beta_c$, and $\beta_p$ based on the first three equations, if it so happens that the average visit duration for (C = 1, P =1 ) is not equal to $\beta_0 + \beta_p + \beta_c$, we're stuck. Our algorithm will do its best to find the values that minimize the error in our regression, but our estimates will be biased. Unfortunately, that's precisely the case we're trying to account for! In other words, simply adding *Children* as a variable in our regression doesn't account for the interaction between *Children* and *PlayArea*. There is no way for us to determine through these equations whether the presence of a play area influences visit duration more for customers with children.

This is where moderation comes in. We can solve our problem by adding a fourth coefficient, for the interaction of *PlayArea* and *Children* (shown in Equation 11-2).

*Equation 11-2.*

$VisitDuration = \beta_0 + \beta_p.PlayArea + \beta_c.Children + \beta_i.(PlayArea * Children)$

The equation for (C = 1, P = 1) becomes *VisitDuration* = $\beta_0 + \beta_p + \beta_c + \beta_i$ and we can adjust the coefficient $\beta_i$ to account for the interaction effect. Now we have:

- If *Children* = 0, the impact of adding a play area is $(\beta_0 + \beta_p) - (\beta_0) = \beta_p$.
- If *Children* = 1, the impact of adding a play area is $(\beta_0 + \beta_c + \beta_p + \beta_i) - (\beta_0 + \beta_c) = \beta_p + \beta_i$.

The difference between these two equations is:

$$[(\beta_0 + \beta_c + \beta_p + \beta_i) - (\beta_0)] - [(\beta_0 + \beta_p + \beta_i) - (\beta_0)] = \beta_p + \beta_i - \beta_p = \beta_i$$

Adding a play area increases visit duration by $\beta_i$ minutes *more* for customers with children than for customers without children.

The multiplication term in Equation 11-2 is traditionally represented in a CD by an arrow ending in the middle of the original arrow (Figure 11-2). In that case, the variable *Children* is called the *moderator*, and the relationship between *PlayArea* and *VisitDuration* is *moderated*.

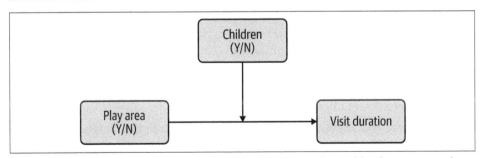

*Figure 11-2. The effect of PlayArea on VisitDuration is moderated by the customers having children with them*

Regression software recognizes moderation, and usually includes a shortcut: if you include only the product of two variables, the software will recognize that you want to calculate coefficients for the individual variables as well:

```
## Python (output not shown)
ols("duration~play_area * children", data=hist_data_df).fit().summary()

## R
> summary(lm(duration~play_area * children, data=hist_data))
...
Coefficients:
                       Estimate Std. Error t value Pr(>|t|)
(Intercept)            19.98760    0.01245  1605.5   <2e-16 ***
play_area1              3.95907    0.02097   188.8   <2e-16 ***
children1              10.01527    0.02017   496.6   <2e-16 ***
play_area1:children1   20.98663    0.03343   627.8   <2e-16 ***
...
```

 When you use the shortcut described in the preceding, your software of choice will add *PlayArea* and *Children* as individual predictors of *VisitDuration*, and estimate Equation 11-2 as a result. Wouldn't it be better to estimate instead the equation without *Children* as a separate term? That is:

$$VisitDuration = \beta_0 + \beta_p.PlayArea + \beta_i.(PlayArea * Children)$$

Short answer: no. For more details, see reference works such as Hayes (2017), but basically for the coefficients to mean what you want them to, include in your regression the moderator and moderated variables as individual variables too, even if their coefficients are not economically or statistically significant. Don't override your software to remove them; this is a feature, not a bug.

The coefficients in this regression match the averages for the four cases we've listed:

- The average visit duration for a customer without children in a store without a play area is $\beta_0$, the coefficient for the intercept, i.e., 20 minutes.

- The average visit duration for a customer without children in a store with a play area is $\beta_0 + \beta_p$, the sum of the coefficients for the intercept and *PlayArea*, i.e., around 20 + 4 = 24 minutes.

- The average visit duration for a customer with children in a store without a play area is $\beta_0 + \beta_c$, the sum of the coefficients for the intercept and *Children*, i.e., around 20 + 10 = 30 minutes.

- The average visit duration for a customer with children in a store with a play area is $\beta_0 + \beta_c + \beta_p + \beta_i$, the sum of the coefficients for the intercept, *PlayArea*,

*Children*, and the interaction term between *PlayArea* and *Children*, around 20 + 4 + 10 + 21 = 55 minutes.

In other words, having a play area has a big impact on the average visit duration of customers with children, and a much smaller but non-negligible impact on the average visit duration of customers without children (the relaxation of shopping tantrum-free, maybe).

This can be represented visually, as in Figure 11-3. The values for the *Children* variable are on the x-axis, and *VisitDuration* is on the y-axis, and we have two lines, one for each possible value of *PlayArea*. In other words, the four points at the extremities of the two lines represent the four cases we just saw, and their y values match the coefficients.

If there was no interaction effect between our two variables, the two lines would be parallel, as having a play area would shift the average visit duration upward by the same increment. The fact that they're not parallel demonstrates that *PlayArea* has a bigger impact on customers with children.

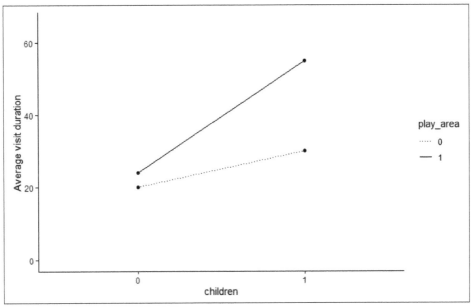

*Figure 11-3. Visual representation of moderation*

If you want to connect this figure back to our math, the distance between the two points on the left (for C = 0) is equal to $\beta_p$ while the distance between the two points on the right (C = 1) is $\beta_p + \beta_i$.

Insights from moderation analysis allow us to better target our actions, e.g., if C-Mart wants to determine in which stores to build play areas next. While the individual coefficient $\beta_p$ is relevant to determine if a play area would pay for itself in general, it is completely irrelevant for store prioritization. Choosing between stores requires determining the store and customer characteristics that have the strongest interactions with the *PlayArea* variable, and then determining which store would yield the highest "lift" from the addition of a play area.

Alternatively, in stores that already have a play area, moderation analysis would allow C-Mart to determine which potential customers to target with a mailing touting the presence of a play area.

### Segmenting experimental data

The process to segment experimental data is pretty much identical to the process we just used with observational data. Therefore I will not repeat the quantitative analyses and I'll just point out some nuances and subtleties to keep in mind.

When running an experiment, we often care not only about measuring the average effect of the treatment in our sample, but also about determining groups for which the effect is particularly strong or weak. Sending a letter or training an employee can be costly; even if financial costs are negligible, as with emails, there might be intangible costs such as annoying a customer. As a result, we generally want to apply a treatment only to the people for which it works.

Beyond cost considerations, personalization is a key rationale for targeting specific messages or treatments to specific segments of our customer base. The very idea of personalized messaging is that individuals outside of the target segment will not react, or may even react negatively to the message. If you have a message that appeals to everyone, good for you, but that's not personalization. For example, a coupon for pro ice-skating blades may annoy the vast majority of people who have no use for it. An ad targeting outdoorsy types may antagonize quiet bookworms and vice versa. Personalization means that you're trading increasing effectiveness on a certain subgroup for decreased effectiveness on another one.

In marketing analytics, this approach is often referred to as "uplift" analysis or modeling, because we're trying to identify the groups of customers for which a given campaign or treatment increases their propensity to take action the most (e.g., buying or voting), *regardless of their initial propensity*. That last point is often a source of confusion so it's worth elaborating: identifying customers with a high propensity can have its uses, but it should not be used as a basis for targeting by itself.

Let's say you compare younger customers (below 30 years old) and older customers (above 60 years old):

- The first group has a probability to take action of 20% if they're not emailed, and 40% if they're emailed.
- The second group has a probability to take action of 80% if they're not emailed, and 90% if they're emailed.

Emailing the first group, the younger customers, would be much more effective on average than emailing the second group, the older customers: it would increase by a larger number the total number of customers taking action.

However, that fact can often be obscured in real life by the lack of a proper control group. If you were to email the second group only and compare their behavior with the rest of the population it would dramatically inflate the apparent effectiveness of the email campaign.

Mathematically, identifying a group with a high effectiveness of treatment means finding demographic variables that are moderators of the effect of the treatment variable on the effect of interest. In that sense, moderation analysis offers us a strong and unified conceptual framework to think about uplift analysis, personalization, and more generally targeting.

## Interactions

The CD we drew to represent segmentation was asymmetrical: the variable *PlayArea* has an arrow going directly to *VisitDuration*, whereas the variable *Children* has an arrow going to the first arrow. However, our regression equation is perfectly symmetrical; nothing indicates which of the two variables is the moderator and which is moderated. Technically speaking, we could interpret Equation 11-2 to mean that *Children* is the cause and *PlayArea* is the moderator.

Is one of these representations more "right" than the other? When we have an individual characteristic and a business or behavioral variable involved in moderation, as is the case here, we generally represent the individual characteristic (e.g., having children) as the moderator, and refer to it as segmentation. That is, the effect of a play area is different between the segment of customers with children and the segment of customers without children.

On the other hand, if we have moderation between two variables of the same type, such as two individual characteristics or two business interventions, it doesn't make sense to introduce an asymmetry between them. For example, we may imagine that having a play area and having a lounge area both increase visit duration independently (the play area for customers with children, and the lounge area for customers without children), but having the two together has a bigger effect than the sum of their individual effects, because customers with children can now use the lounge area.

In such circumstances, we can represent moderation by joining arrows together as in Figure 11-4, and refer to it as *interaction*.

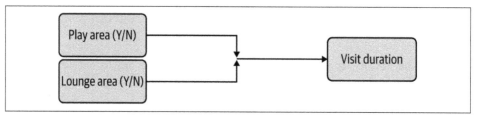

*Figure 11-4. Representing symmetrical interactions*

The equation for this CD would be:

$$VisitDuration = \beta_0 + \beta_p.PlayArea + \beta_l.LoungeArea + \beta_{pl}.(PlayArea * LoungeArea)$$

As you can see, the equation is exactly the same as for segmentation, and if we wanted we could call *LoungeArea* the moderator of the effect of *PlayArea* on *VisitDuration* (or vice versa). Conceptually, interactions let us think about situations where "the whole is more than the sum of the parts," such as complementary behaviors.

 On a side note, it is my belief that some of the power of newer and more complex machine learning methods like random forests, XGBoost, or neural networks comes from their ability to capture such interactions. By including interactions in regression, we can reduce the performance gap while preserving the interpretability of causal relationships we have with regression.

## Nonlinearities

In many circumstances, the relationship between a cause and an effect is not linear. It may have what economists call "decreasing returns": you get less and less bang for your buck. For example, a satisfied customer might buy more than an unsatisfied one, but an ecstatic customer might not buy that much more than a happy one. Sending one marketing email per month to a customer might increase purchases, but sending 11 emails instead of 10 probably doesn't help much, as in the left panel of Figure 11-5.

Conversely, a causal relationship may have increasing returns, such as the network effects touted by start-ups: the more properties AirCnC has on its website, the more customers it will attract; but then the wider its customer base, the higher the incentive for owners to offer their property on its website, and so on, as in the right panel of Figure 11-5.

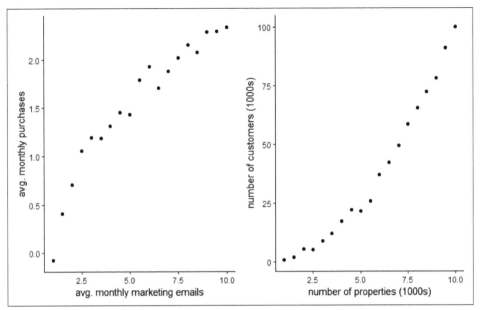

*Figure 11-5. Nonlinear relationships between variables: left is decreasing returns while right is increasing returns*

 Mathematically, the curve on the left is concave, and the curve on the right is convex. A common mistake is to believe that such relationships cannot be represented with linear regression. In reality, what matters for a regression to be "linear" is that the predicted variable has a linear relationship with the coefficients, not the variables. $Y = \beta_1.e^{X_1} + \beta_2.e^{X_2}$ is a proper linear regression because multiplying each coefficient by 2 would multiply Y by 2. Conversely, $Y = e^{\beta_1.X_1} + \beta_2 X_2$ is not a proper linear regression because Y does not have a linear relationship with the coefficients.

We can address nonlinear relationships between variables by adding the explanatory variable taken to the square (i.e., a quadratic term). For example, we can model the relationship between marketing emails and purchases we just discussed as:

*Purchases* = $\beta_0 + \beta_1.$*Emails* + $\beta_2.$*Emails*$^2$

Adding a quadratic term can significantly improve the accuracy of a regression. In Figure 11-6, you can see that the solid curve, which represents the line of best fit for the linear regression with a quadratic term, is much closer to the data points than the dashed line, which represents a standard regression without a quadratic term. Here, additional monthly emails have a decreasing impact, which translates into the quadratic term having a negative coefficient in the regression.

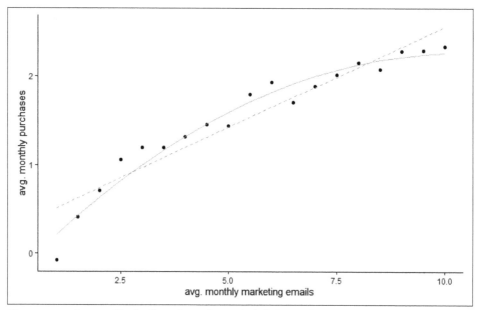

*Figure 11-6. Linear (dashed) and quadratic (solid) lines of best fit*

However, a quadratic term is nothing more than an interaction between a variable and itself. In other words, a nonlinear causal relationship between two variables can be reframed as self-moderation (Figure 11-7).

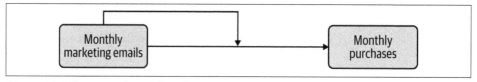

*Figure 11-7. Representing self-moderation*

Conceptually, this means that adding an extra email per month has a different impact depending on the current business-as-usual number of emails sent.

The syntax in code to include self-moderation is:

```
## R
summary(lm(Purchases ~ Emails + I(Emails^2), data=dat))
```

```
## Python
model = ols("Purchases ~ Emails + I(Emails**2)", data=dat_df)
print(model.fit().summary())
```

In both R and Python, this is done by using the identity function I( ), which prevents the linear regression algorithm from trying to interpret the squared term and just passes it to the general solver. In R, a squared term is expressed with a caret,[1] whereas in Python it is expressed with two multiplication symbols. This also implies that self-moderation is validated in exactly the same way as traditional moderation: we would build a Bootstrap confidence interval, and determine whether it includes zero and whether it is economically significant.

To recap our exploration of moderation, we've seen three varieties of it, which are mathematically identical but have different interpretations based on the types of variables involved:

*Segmentation*
> In segmentation personal characteristics (e.g., demographic variables) moderate the effect of a business behavior, such as an experimental intervention (it is then known as uplift analysis).

*Interaction*
> In interaction we observe moderation between variables of the same nature, such as two demographic or two behavioral variables.

*Nonlinearities*
> With nonlinearities a variable self-moderates its causal impact on another variable.

Now that we have a clear understanding of the behavioral interpretation of moderation, let's turn to the details of its application.

# How to Apply Moderation

As we discussed in the previous section, moderation can be used to capture a variety of behavioral effects, simply by adding the product of two variables to a regression. In this section, we'll turn to technical considerations:

- When to look for moderation
- How to validate it
- Moderated moderation
- How to interpret the coefficients for individual variables in a moderated regression

---

1 The caret is the symbol on the 6 key of your keyboard.

# When to Look for Moderation?

With so many possible applications of moderation, it can be tempting to look for it everywhere, but as a second-order effect (i.e., an effect on an effect), moderation generally yields small coefficients and the risk of false positives is high. This is especially true with experimental data, where the motivation is strong to find a meaningful effect: "Sure, the average effect of the email campaign is almost null, but look at the response rate for thirty-year-old men in Kansas!"

Let's say that you're getting started on analyzing observational data or designing an experiment. At what point should you be thinking about moderation, and how should you integrate it into your analyses? First, I'll talk about the experimental design stage. The process for the data analysis stage is the same regardless of whether the data is observational or experimental in nature, so I'll cover the two cases together after that. Finally, I'll cover nonlinearities, which will be easy: because there is only one variable involved, you don't really have to worry about potential risks and you can include nonlinearity liberally in your analyses.

## Including moderation in the experimental design stage

I'll distinguish between two situations that will call for different approaches:

- The primary object of your analysis is the main effect, regardless of moderation, and moderation is a secondary object of your analysis.
- Moderation is the primary object of your analysis.

If you're designing an experiment in which the primary goal is not to measure moderation, my recommendation will be simple: leverage the possibility of moderation to refine your theory of change, but do not try to adjust your sample size.

In Part IV, we saw that before running an experiment, you should make your theory of change explicit with the help of causal diagrams. In that case, the focus was on average causal effects, i.e., the average effect of an experimental treatment across the board, but in many cases, we can refine that logic through moderation.

In Chapter 8, we designed an experiment where we sought to increase AirCnC's booking rate by offering "1-click booking." It is conceivable that this effect could be moderated by age (Figure 11-8).

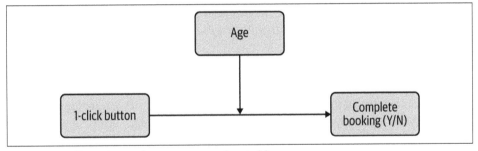

*Figure 11-8. Experimental treatment moderated by age*

The behavioral logic of our theory of change was that the 1-click button would reduce the duration of the booking process, which itself impacts the probability of completing a booking. This means there are two possibilities in terms of moderation: age moderates the effect of the 1-click button on the booking duration, or it moderates the effect of booking duration on the probability of completing a booking, or both (Figure 11-9).

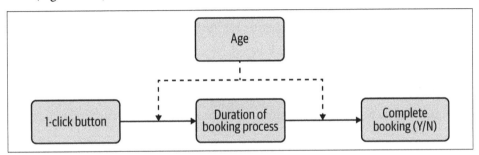

*Figure 11-9. Age may moderate two causal relationships*

Putting together our behavioral logic and a potential moderator is very powerful, because it allows us to think through the behavioral implications, and in some cases to run analyses, before our experiment takes place.

Let's start with the first relationship on the left of the CD, between the 1-click button and the booking duration. We don't have any data about the 1-click button in our historical data, so we can't measure this moderation effect directly. However, if we see an (unconfounded) correlation in our historical data between age and booking duration, this lends some credence to the idea, because it suggests that the cognitive processes at play are affected by age. On the other hand, if age is not correlated with booking duration, moderation would rely on younger and older customers having different reactions to the button itself specifically. While not impossible, this is a "narrower" behavioral path, somewhat less likely per Occam's Razor—the simplest explanation is generally the right one. More importantly, it is a testable behavioral path. For example, if you bring a sample of baby boomers into the UX lab and you find that they don't trust a 1-click process because they want to go through all the steps, then moderation is very likely, and you can target your experiment to younger customers, or at least oversample them.

For the second relationship, between booking duration and booking completion, we have all the required variables in our historical data, which means we can confirm or disprove moderation before even running our experiment, with a much higher accuracy than with the limited sample size of a single experiment. Again, if we confirm the presence of moderation, we can adjust our experimental design accordingly (e.g., by targeting only younger or older customers, depending on the direction of moderation).

There is a broader takeaway here: by articulating the behavioral logic of an intervention, we can often expand the chain between our intervention and our effect of interest in the CD by identifying one or several mediators between them for which we already have data. We can then explore whether the relationship between mediator(s) and final effect is moderated. In addition, we can explore whether that relationship is self-moderated. In our example here, maybe the booking rate is unaffected by the booking duration up to a point, but then drops precipitously, e.g., customers don't really care if booking takes 30 or 45 seconds, but bail out of the process en masse if it takes more than 2 minutes (Figure 11-10).

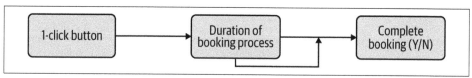

*Figure 11-10. Relationship between the mediator and final effect is self-moderated*

 During that process of identifying moderators in your historical data, you'll have to keep in mind that "correlation is not causation" applies, and the relationships you're looking at might be confounded, as in Figure 11-11.

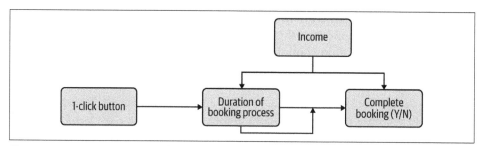

*Figure 11-11. Income is a confounder of the relationship between duration and booking rate*

In such a situation, the confounding effect of *Income* would bias the relationship between *Duration* and *Booking* compared to the real causal effect (which is what you would act upon through the 1-click button experiment). Therefore, you would need to control for the confounding effect when assessing moderation.

In the case of a variable like *Age* that is observable before our experimental allocation, we can use it directly for the allocation itself. For example, instead of drawing a random sample from our whole customer base, we can limit our experiment to customers under a certain age, where we expect the largest effect.

Obviously, we can't observe *BookingDuration* before a customer starts the booking process, which means we can't use it in the experimental allocation itself. However, we can try to identify proxies for the target variable: for example, maybe bookings tend to be longer on the weekend than on weekdays. We can then use these proxies to define our experimental population and, say, run our experiment only on weekend customers. This will work well if the proxy and the target variable are closely related, e.g., there are some shorter bookings as well over the weekend, but 90% of the longer-duration bookings happen over the weekend. Conversely, if the two variables are only loosely related, e.g., 60% of the longer-duration bookings happen over the weekend, the results observed in your experiment may not generalize well to longer-duration bookings on weekdays.

This is all I recommend you do if moderation is not the primary focus of your experiment, i.e., you haven't measured the unmoderated effect yet. That's because the power to detect moderation and interactions is significantly lower than the power to detect the main effects, which means the sample size required to measure moderation would be significantly larger (think 10–20x larger or more). That's a lot of time spent on the possibility of moderation for an average effect you haven't even confirmed yet.

Therefore, I recommend that you run your first experiment with a focus on measuring accurately the average, unmoderated effect, and then measure moderation effects in the results. If you find any promising moderation effect, it will most likely have a large confidence interval that crosses zero, due to the fact that the experiment was not powered to capture it. If that moderation effect seems promising enough from an economic standpoint, run a second experiment powered to appropriately measure it.

Once you know the unmoderated effect size, you can decide to run an experiment dedicated to measuring or confirming moderation. The process for determining the appropriate sample size remains the same: you would set a target power and hypothesized/target effect sizes, then determine the proportion of false negatives in repeated simulations at different sample sizes, as we've done in previous power analyses. The only difference is that you would use your prior knowledge for the main effect size, and only set a target effect size for moderation.

## Including moderation in the data analysis stage

You have some data on hand, either observational in nature or from an experiment you have run. How can you identify relevant moderation effects without running into false positives? You'll have to do a fishing expedition, that is, try including a moderation term with one variable, then another, and so on. In the following, I'll provide several guidelines to minimize the risk of false positives.

First, I'll provide you with a rough but powerful sanity check when you're looking to moderate the effect of a categorical variable on a numeric variable. The requirement of a numeric effect simply means that you're looking at a linear and not a logistic regression. The requirement of a categorical cause may seem very restrictive, but it applies to all experimental assignments (i.e., the experimental assignment is always a binary or categorical variable). In addition, if your cause is numeric, you can discretize it for this purpose by taking its quartiles:

```R
## R
hist_data <- hist_data %>% mutate(age_quart = ntile(age, 4))

## Python
hist_data_df['age_quart'] = pd.cut(hist_data_df['age'], 4,
                                    labels=['q4', 'q3', 'q2', 'q1'],
                                    include_lowest=True)
```

The sanity check is to compare the standard deviation of the effect of interest across the groups defined by your cause of interest. If the standard deviation is meaningfully higher in the treatment group (for experimental data) or differs across groups (for observational data), this suggests the possible presence of moderation and you can proceed with your fishing expedition with reasonable confidence. If the standard deviations are similar across groups, this suggests that there is no moderation; you can still try a few potential moderators if you'd like, but you need to have a strong theoretical rationale for them.

What do "meaningfully higher" or "similar" mean in that context? If you want a rigorous justification, there are statistical tests you can run to help you determine if the observed difference is statistically unusual, such as the Brown–Forsythe test.[2] Personally, I would recommend simply eyeballing it: is the difference economically meaningful in relation to the difference in means across the groups?

---

2 Available in R as the `bf.test()` function from the `onewaytests` package (*https://oreil.ly/iq7KN*) and in Python as the `stats.levene()` function from the `scipy` package, with the parameter `center='median'`.

Going back to the example of play areas in C-Mart stores, the code would be as follows:

```
## R (output not shown)
> hist_data %>% group_by(play_area) %>% summarize(mean = mean(duration),
                                                  sd = sd(duration))

## Python
 hist_data_df.groupby('play_area').agg(M = ('duration', lambda x: x.mean()),
    SD = ('duration', lambda x: x.std()))
Out[22]:
                 M         SD
play_area
0           23.803928   6.970786
1           36.360939  17.111469
```

In the example here, a 10-minute difference in standard deviations is definitely something C-Mart would be interested in exploring, given that the difference in means across the groups is about 13 minutes.

 Obviously, the risk of false positives increases with the number of categories. If your primary cause is state or profession, you're likely to see some variations in standard deviations across the board, just out of randomness and special cases. Therefore, I would look at moderation for such a variable only if I had a pretty strong rationale in the first place.

One way to mitigate the risk of false positives and make any moderation you might find more meaningful is to replace your categorical variable by quartiles of a relevant numeric variable. For example, political leaning or average income for states, proportion of women or average level of education for profession. Saying that the 25% of states with the lowest average income have a lower standard deviation of purchases than the 25% of states with the highest average income is a much more robust and actionable insight than saying that California has a higher standard deviation than Mississippi.

Assuming your data passes this first sanity check, the second step is to establish upper bounds on moderated effects. The key intuition here is that moderation can only "redistribute" the average effect; it doesn't increase it. A visual illustration will make this clearer, taking the example of *Children* as a potential moderator of the effect of *PlayArea* on *VisitDuration*. Figure 11-12 shows the average effect of a play area across our data (11.92 minutes), with the width of the bars representing the proportion of customers without and with children in our population.

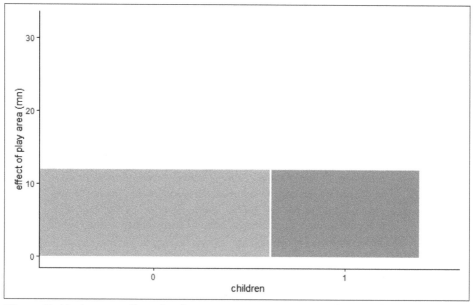

*Figure 11-12. Average effect of a play area across customers without and with children*

Let's assume for a second that the presence of a play area has at worst a null effect on customers without children (i.e., it cannot have a negative effect), a plausible assumption from a behavioral perspective if the play area is appropriately soundproofed. This implies that, at most, the entire average effect is coming from customers with children, and moderation redistributes the entire area of the left bar to the right bar, as in Figure 11-13.

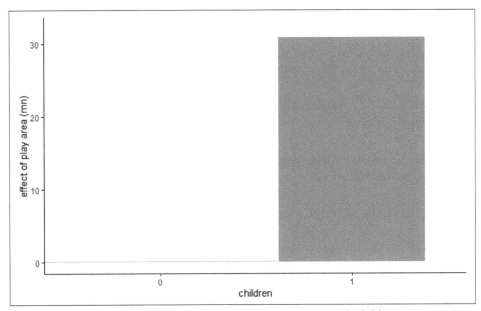

*Figure 11-13. Average effect coming entirely from customers with children*

Under that "most extreme scenario," let's calculate the effect size in the group of customers with children. By definition of the average, we have:

*Avg. effect size = (Sum of individual effect sizes) / (Total number of customers)*

Let's separate customers with and without children in the numerator:

*Avg. effect size = (Sum of individual effect sizes for customers without children + sum of individual effect sizes for customer with children) / (Total number of customers)*

If we assume no effect on customers without children, the first term in the numerator is equal to zero:

*Avg. effect size = (Sum of individual effect sizes for customers with children) / (Total number of customers)*

Let's multiply both sides of the equation by the total number of customers:

*Avg. effect size * Total number of customers = Sum of individual effect sizes for customers with children*

And then divide both sides by the number of customers with children:

*Avg. effect size \* Total number of customers / Number of customers with children = Avg. effect size for customers with children*

*Avg. effect size \* 1 / Proportion of customers with children = Avg. effect size for customers with children*

Swapping left- and right-hand sides for readability and plugging in the numbers from the preceding example, we get:

*Avg. effect size for customers with children = 11.92 \* 1 / 0.387 = 30.8*

In other words, the average effect size for customers with children can be no higher than 30.8 additional minutes for *VisitDuration*, given the overall average effect size. If that number is too low to be of economic interest, then there is no point in measuring moderation between *PlayArea* and *Children*. Beyond that specific example, the formula *Avg. effect size/Proportion of customers in segment* can be applied to any potential moderator: if you have an equal split of men and women in your customer base, this means that at most, one gender has double the average effect and the other has no effect. Hoping that somehow one gender will have triple the average effect or that moderation will increase the effect for both genders is wishful thinking, and mathematically impossible. To put it another way, we tend to focus on groups with above-average effect sizes, for understandable reasons, but the counterpart has to be groups with below-average effect sizes.

This means that your search for potential moderators should start with variables that have a strong behavioral rationale or have been found to be impactful in past analyses, and that generate subgroups large enough to matter. Beyond a certain degree of inequality in group sizes, the search becomes an exercise in futility: if you have a variable that splits your customer base 90%-10%, then even if the group representing 10% of your customer base shows no effect at all, this will increase at most the effect in the 90% group to 1 / 0.9 = 111% of the average effect, an increase of 11%.

More broadly, moderation cannot save a mediocre average effect. It should only be sought to add a bit more oomph to your treatment; for example if you are at 90% of the breakeven or target value for it, or if you are in a high-volume environment and you're trying to extract any efficiency gains you can.

For each variable you want to test, the process is the same. First, run a regression with an interaction between the experimental treatment variable and the potential moderator. Then if you find an interaction effect large enough, confirm it by building a Bootstrap CI around the estimated effect.

There are some rules, such as the Bonferroni correction, to reduce the risk of false positives when testing a large number of hypotheses in this fashion. I would not recommend them however, for two reasons:

1. They generally rely, explicitly or implicitly, on the Null Hypothesis Statistical Testing framework with normality assumptions.

2. They can be unduly conservative and unacceptably increase the risk of false negatives.

Instead, my recommendation would be to validate promising subgroups through follow-up experiments whenever possible. If a moderation effect is large enough to be of business value, then it should be considered large enough to warrant further validation. If you're working with experimental data, repeating the experiment is conceptually straightforward. If you're working with observational data, what experiment you should run may not be immediately clear. However, for your moderation effect to have any economic value, it must mean you're planning to do something differently. Otherwise it's just an interesting tidbit for cocktail parties. Whatever it is you would do differently, you can probably randomize it in some fashion, and now you have an experiment.

One key success factor in this process is to appropriately convey to business partners that the results of such repeated testing should be considered provisional hypotheses, not proven fact. They shouldn't feel like you've failed to deliver when an experiment yields a null result because it was a long shot anyway. Also remember that due to the nature of the process, your final effect size is likely to be smaller than what you found in the first pass.

## Nonlinearities

Nonlinearities, a.k.a. self-moderation, represent a special case compared to other forms of moderation because, by definition, there is only one moderator under consideration, which limits the risk of false positives. In addition, the consequences of a false positive are typically limited, as long as you don't draw inference outside of the range of data available. For example, if you were to measure the effect of income on purchases based mostly on customers with annual incomes between $25,000 and $75,000 and then draw inferences for customers with an annual income of $250,000, both the risks and consequences of a false positive are high (even without considering moderation, extrapolating that far out would be a terrible idea anyway).

Therefore, it's mostly OK to routinely test your cause of interest for self-moderation if you have at least a few hundred rows in your data. The variable must also be numeric, because a categorical variable self-moderating would be nonsensical. You should then validate the self-moderation effect by building a Bootstrap confidence interval, as we'll see a bit later in this section.

Beyond improving the fit of your regression and accounting for intuitive behavioral effects such as decreasing returns, including self-moderation in your regression can alert you to the presence of a hidden moderator.

Let's look at a C-Mart example: the relationship between visit duration and groceries purchases. It is conceivable that very short visits represent targeted shopping runs to buy a specific article whereas longer trips are more likely to be grocery runs. This would make the reason for a visit a confounder of the relationship between *VisitDuration* and *GroceriesPurchases* (Figure 11-14).

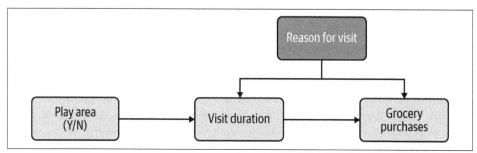

*Figure 11-14. Reason for visit is a confounder of the relationship between VisitDuration and GroceriesPurchases*

But at the same time, *VisitReason* might conceivably moderate the effect of *VisitDuration* on *GroceriesPurchases*. If you're going to the store to grab a birthday gift or a hammer, keeping you in the store longer (e.g., by adding a play area) is unlikely to entice you to buy sun-dried tomatoes the way it might if you're doing a grocery run (Figure 11-15).

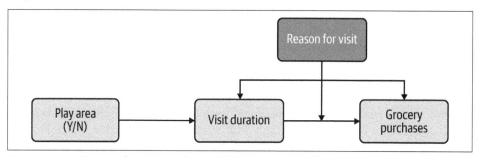

*Figure 11-15. Reason for visit moderates VisitDuration's effect on GroceriesPurchases*

If we were able to observe the reason for the visit, we would be able to add it to our CD, and it would become a straightforward case of confounding + moderation. Assuming that we can't observe the reason for the visit, we're left with the following observable facts:

- *GroceriesPurchases* is positively correlated with *VisitDuration* (both because of a true effect and because of the confounding effect of reason for visit).
- Increasing visit duration has a bigger impact on groceries purchases for longer visits than for shorter visits.

In other words, the relationship between *VisitDuration* and *GroceriesPurchases* exhibits nonlinearity. A self-moderation term would increase the accuracy of our regression, so we should include it.

More generally, whenever you have a variable that appears to self-moderate without a clear behavioral rationale, you should explore the possibility of a hidden variable that is both a cause and a moderator of that variable.

To recap, finding relevant moderators is an important part of behavioral analysis, but like most of our tools, it is often more art than science. In the experimental design stage, we can sharpen our behavioral logic by looking for moderated mediators in our historical data, or through UX testing of our treatment. In the data analysis stage, a fishing expedition can discover promising potential moderators; while Bootstrap confidence intervals help reduce the risk of false positives, follow-up experiments are ultimately your best guarantee of success. Conversely, self-moderation is safe to explore and include routinely.

# Multiple Moderators

For the sake of simplicity, so far I've only shown one moderator at a time, but an effect can have multiple moderators. Going from one to multiple moderators is straightforward; the main subtlety to keep in mind is whether the moderators are interacting with each other or not.

### Parallel moderators

Getting back to our C-Mart example, we could imagine that another demographic variable, *Age*, also separately moderates the effect of *PlayArea* on *VisitDuration*, as in Figure 11-16.

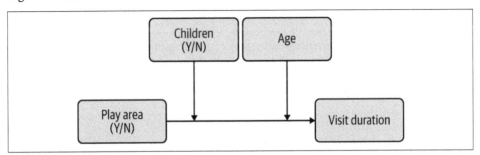

*Figure 11-16. Effect of PlayArea on VisitDuration has two moderators*

The corresponding regression equation is shown in Equation 11-3.

*Equation 11-3.*

$VisitDuration = \beta_0 + \beta_p.PlayArea + \beta_c.Children + \beta_a.Age + \beta_{pc}.(PlayArea * Children) + \beta_{pa}.(PlayArea * Age)$

Note that the equation includes an individual term for *Age*, even though there is no arrow going from *Age* to *VisitDuration*. The warning provided earlier still applies here, and you need to include all individual terms in your regression.

The interpretation of Equation 11-3 would be that:

- The impact of *PlayArea* on *VisitDuration* is different for customers with and without children.
- The impact of *PlayArea* on *VisitDuration* is different for younger and older customers. These two moderating effects are independent of each other.

*Age* is a numeric variable, so its interpretation must be adjusted accordingly: the coefficient for *PlayArea * Age* represents the difference in the effect of *PlayArea* on *VisitDuration* between customers who have a one-year age difference. Depending on the business problem at hand, we can either:

- Keep it in numeric format if we wanted to have a precise, causally grounded estimate of *VisitDuration*, for example to determine the increase in sales we should expect from adding a play area to a certain store.
- Or make it categorical by "binning" it in an appropriate manner. For example, we could convert *Age* into brackets such as "less than 20," "20 to 40," "40+," or any other breakdown, making the corresponding coefficients easier to interpret for segmentation purposes.

Taken together, *Children* and *Age* create a two-dimension demographic segmentation that allows us to compare the average visit duration of, say, a 28-year-old with children to a 45-year-old without children.

Because the two moderation effects are independent of each other, we would be able to validate each one of them independently by bootstrapping the regression for Equation 11-3 and looking at the confidence interval for each moderator (see the next subsection).

This also implies that there is no order between the two moderators, and they can be drawn interchangeably: in Figure 11-16, *Children* comes first but *Age* could as well. It doesn't matter.

The logic of multiple moderators applies similarly for interactions between variables of the same nature. For example, we could have *Children* interact with both *Age* and *Gender* (Figure 11-17).

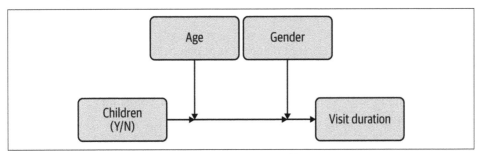

*Figure 11-17. Age and Gender both interact with Children*

The corresponding regression equation would be:

$VisitDuration = \beta_0 + \beta_c.Children + \beta_a.Age + \beta_g.Gender + \beta_{ca}.(Children * Age) + \beta_{cg}.(Children * Gender)$

Finally, a regression can include multiple variables that each self-moderate. Here again, the analysis would proceed independently for each one of them.

Overall, it is very simple to add multiple independent moderators, but sometimes it also makes sense to assume that the moderators interact with each other, which we'll now turn to.

### Interacting moderators

Looking at the effect of *PlayArea* on *VisitDuration*, it makes sense that it would be moderated both by *Children* and *Age*. But among customers with children, we may also imagine that the increase in visit duration depends on the age of the customer, e.g., if grandparents are less likely to drop off their grandchildren at the play area than parents are to drop off their children. In such a case, the moderating effect of *Children* on the effect of *PlayArea* on *VisitDuration* would itself be moderated by *Age*, which in social sciences is called "moderated moderation" (Figure 11-18).

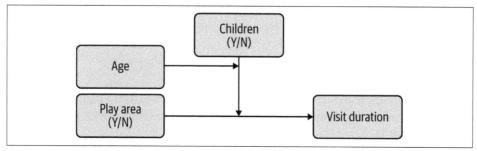

Figure 11-18. Moderated moderation

Figure 11-19 shows the moderated moderation at play, with each subgraph represent-ing an age group among our customers. We can see that going from younger to older customers, the impact of *Children* on the impact of *PlayArea* on *VisitDuration* decrea-ses (quite a mouthful!), as the distance between the two points for C = 1 decreases across subgraphs.

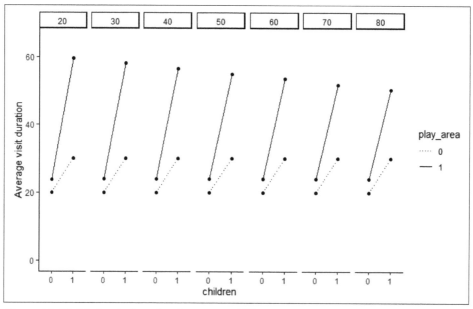

Figure 11-19. Moderated moderation across different age groups

The corresponding equation would be as shown:

$$VisitDuration = \beta_0 + \beta_p.PlayArea + \beta_c.Children + \beta_a.Age + \beta_{pc}.(PlayArea * Children)$$
$$+ \beta_{pa}.(PlayArea * Age) + \beta_{ca}.(Children * Age) + \beta_{pca}.(PlayArea * Children * Age)$$

This equation is identical to Equation 11-3 with the addition of a three-way term at the end. Let's run the regression (note how I entered only the three-way interaction term in the regression and the software automatically adds all the individual variables and two-way interaction terms):

```
## R (output not shown)
> summary(lm(duration~play_area * children * age, data=hist_data))

## Python
ols("duration~play_area * children * age", data=hist_data_df).fit().summary()
```

```
...
                          coef   std err     t     P>|t|    [0.025   0.975]
-----------------------------------------------------------------------------
Intercept              20.0166    0.037  534.906   0.000   19.943   20.090
play_area               3.9110    0.063   62.014   0.000    3.787    4.035
children                9.9983    0.061  165.012   0.000    9.880   10.117
play_area:children     29.1638    0.101  290.105   0.000   28.967   29.361
age                    -0.0006    0.001   -0.820   0.412   -0.002    0.001
play_area:age           0.0010    0.001    0.806   0.420   -0.001    0.003
children:age            0.0003    0.001    0.297   0.767   -0.002    0.003
play_area:children:age -0.1637    0.002  -86.139   0.000   -0.167   -0.160
...
```

The coefficient for our 3-way term, the last one in the output, is negative, as well as economically meaningful, and the 90%-CI is approximately [–0.1671 ; –0.1590], which is narrow enough to give us confidence it can't be close to zero.

Moderated moderation obeys the same logic and rules as simple moderation. Therefore it is symmetric, meaning that we can interpret *Age* as moderating the moderating effect of *Children*, or we can see it as *Children* moderating the moderating effect of *Age* the last one in the output.

From a behavioral perspective, the interpretation of moderated moderation depends on whether the underlying logic is that of segmentation or interaction:

- We would interpret it as segmentation if we have two personal-characteristic variables moderating a business characteristic or business behavior. That was the case for instance with the play area, where *Children* and *Age* moderate the effect of the *PlayArea* variable. Intuitively, this means that we have a two-dimension segmentation where the moderating effect of a dimension is increasing or decreasing along the other dimension (e.g., the moderating effect of having children decreases when *Age* increases).

- With three variables of the same nature, we would interpret it as a three-way interaction between three variables, as in Figure 11-20.

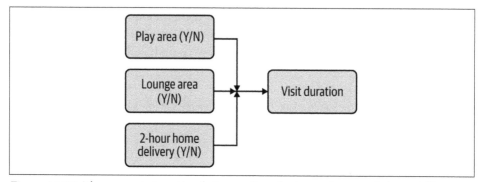

*Figure 11-20. Three-way interaction between three business-characteristic variables*

The point of this distinction is not to be pedantic, but to remind you to keep track of your analysis goals. Finding and validating moderation between predictors in a regression is all well and good, but what are you going to do with that information? You can *change* business behaviors, and sometimes business characteristics, but you can only *target* customers' personal characteristics. Similarly, if you find that a certain customer behavior moderates the effect of a business behavior (e.g., a cross-selling email campaign works best on customers who have recently been in a store), that information could be used either to target the customers who have exhibited this behavior, or to encourage customers to have that behavior in the first place.

Technically speaking, moderated moderation could also apply to nonlinearities, by having a cube term (e.g., *Purchases* = $\beta_0$ + $\beta_1$.*Emails* + $\beta_2$.*Emails*$^2$ + $\beta_3$.*Emails*$^3$), but that's extremely rare and more of a curiosity devoid of practical applications.

At this point, you may have a budding or even full-fledged headache. Moderation can quickly get overly complex. I have shown you 3-way interactions to let you know that it's a possibility, but I would not go down that path unless you have some reasonably strong preexisting behavioral rationale to do so. Beyond that, we could theoretically have 5-way or even 12-way interactions ("moderation of moderation of..."), but simply throwing in as many interaction terms as you can is a recipe for false positives and ultimately a dazzling but pointless model. A straightforward simple moderation is often all you need, and can add some significant oomph to your analysis at a reasonable cost in terms of added complexity.

# Validating Moderation with Bootstrap

So far, we have only looked at the regression coefficients for moderation terms, and not at p-values. But like any other regression coefficient, our coefficients for moderation are subject to uncertainty and sampling variability. Uncertainty is all the more important to account for with moderation because it is a "second order" effect (i.e., an effect on an effect, not a direct effect on a variable); these are typically much smaller than "first order" effects.

Let's get back to our example of play areas in C-Mart. Our estimated coefficient for the moderation between *PlayArea* and *Children* was $\beta_{pc} = 21$. As usual, we'll use Bootstrap simulations to determine how certain we can be of the values we observed. Let's start by drawing 1,000 random samples of 10,000 rows each from the historical data available to us, and running each time the same regression as in the previous section. The distribution of values for the interaction coefficient is displayed in Figure 11-21.

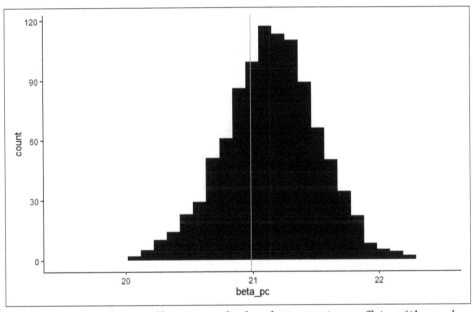

*Figure 11-21. Distribution of bootstrapped values for interaction coefficient (1k samples of 10k rows)*

In Chapter 7, we saw that increasing the number of samples increased the smoothness of the histogram and the accuracy of confidence interval estimates, and increasing the size of the samples reduced the dispersion of values. Whether we need to do either of these depends on the business question at hand:

- If the question is, "Is there any moderation whatsoever?" (i.e., Is the coefficient $\beta_{pc}$ different from zero?), then Figure 11-21 already allows us to answer unequivocally "yes" and there's no need to dig deeper.

- If the question is more uncertain, e.g., "How confident are we that the coefficient for moderation is above 20.5?" then we'll have to increase the sample size. Right now, the histogram goes beyond 20.5 toward the left (and quite so). Each of the values lower than 20.5 represents a Bootstrap simulation where the coefficient for moderation was below that threshold. Let's next increase the size of the Bootstrap samples to 200,000 rows (Figure 11-22).

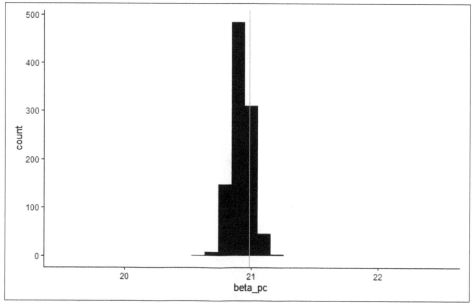

*Figure 11-22. Distribution of bootstrapped values for interaction coefficient (1k samples of 200k rows)*

As you can see, the values are now much more narrowly concentrated around 21, and safely away from 20.5. That second simulation was enough to answer our business question.

In the present case, we have a grand total of about 600,000 rows in our historical data so after validating the code with a dry run, we could have jumped directly to that sample size without much inconvenience (remember that you should never use a Bootstrap sample size larger than the size of the data you're drawing from). But what if your historical data had 10 or 100 million rows? Instead of answering your business question in a matter of minutes, you would have spent hours or days waiting for the results of simulating a thousand samples of that size; certainly, your final CI would be extremely narrow, for example [20.9999; 21.0001], but that would be a complete waste of time if the question is only, "Is it higher than 20.5?" That's why I wanted to show you the process of progressively increasing the Bootstrap sample size, instead of jumping directly to the size of your historical data.

To summarize, after running a small Bootstrap simulation to make sure that your code runs correctly, you should increase the number of samples or the sample size as needed to answer your business question. This iterative process has the added benefit of engaging your critical sense and preventing you from running the analysis on autopilot. This applies to all forms of moderation: segmentation (including of experimental data), interaction, and self-moderation.

## Interpreting Individual Coefficients

I mentioned several times in this chapter that the variables involved in moderation must also be included as individual variables in the regression, even if you don't plan to use them or they don't appear significant. If, however, you want to use them, there are some subtleties involved, that we'll now review.

Let's start by comparing the following two regressions (including only *Age* beyond *PlayArea* for the sake of simplicity):

$$VisitDuration = \beta_0 + \beta_{p0}.PlayArea + \beta_{a0}.Age$$

and

$$VisitDuration = \beta_1 + \beta_{p1}.PlayArea + \beta_{a1}.Age + \beta_{pa1}.(PlayArea * Age)$$

At first glance, it might seem that the interpretation of the second equation is identical to the first one, with the only addition of the moderation term. It is not. $\beta_{p0}$ and $\beta_{p1}$ are not equal and do not have the same meaning, and similarly for $\beta_{a0}$ and $\beta_{a1}$.

Let's visualize the difference by drawing a sample of 1,000 data points for the sake of readability and representing them in the *Age * VisitDuration* plane (Figures 11-23 and 11-24). In both figures, there are two regression lines plotted: one for customers with a play area and one for customers with no play area.

Figure 11-23 represents the first equation, without moderation. The two lines have the same slope, equal to $\beta_{a0} = -0.024$, and the (constant) distance between the two lines is equal to $\beta_{p0} = 12.56$.

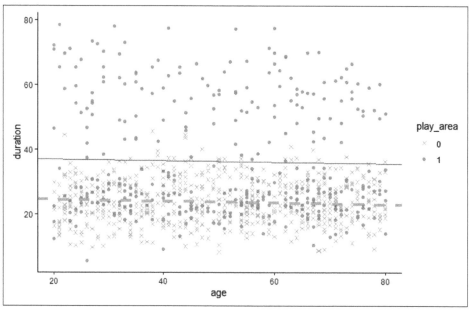

*Figure 11-23. Sample of 1,000 data points and regression lines with a play area (full dots, solid line) and without (crosses, dashed line), without moderation term*

Figure 11-24 shows the corresponding figure for the second equation. The lines are not parallel. This implies that we can't just say, "$\beta_{p1}$ represents the vertical distance between two regression lines," we have to state at which age we're measuring that vertical distance. Similarly, we can't just say that "$\beta_{a1}$ represents the slope of these regression lines," we have to pick which of these lines we're referring to (i.e., we have to state for which value of *PlayArea* we're measuring that slope).

This has important ramifications from a business perspective. If a business partner asks, "What is the impact of a play area on visit duration?" and you're relying on the first equation, without a moderation term, you can answer, "That impact is equal to $\beta_{p0}$." However, if you've determined that there is a significant moderation effect between the two variables and you want to rely on the second equation (as you should), your answer should really be the "well, it depends" that business partners dread.

Fortunately, this problem can easily be resolved with a little care, in two possible ways:

- Setting meaningful reference points
- Calculating effects at the level of business decisions

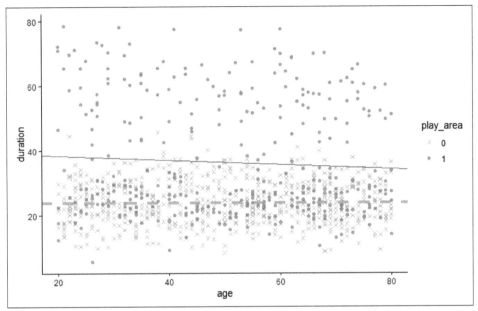

*Figure 11-24. Sample of 1,000 data points and regression lines with a play area (full dots, solid line) and without (crosses, dashed line), with moderation term*

## Setting meaningful reference points

The first solution is to set meaningful reference points for our variables. First let's get back to our equation with moderation:

$$VisitDuration = \beta_1 + \beta_{p1}.PlayArea + \beta_{a1}.Age + \beta_{pa1}.(PlayArea * Age)$$

If we set a customer's age to zero, the expected visit duration for our newborn customer is $\beta_1$ without a play area and $\beta_1 + \beta_{p1}$ with a play area. Therefore $\beta_{p1}$ is truthfully the impact of a play area on visit duration for age zero. However, newborn customers are not our primary target and we can do better by normalizing the *Age* variable. A typical approach is to set it to the average age in our data set:

```
## R
> centered_data <- hist_data %>% mutate(age = age - mean(age))

## Python
centered_data_df = hist_data_df.copy()
centered_data_df['age'] = centered_data_df['age'] \
    .subtract(centered_data_df['age'].mean())
```

Doing so reduces the coefficient for *PlayArea* from 15.85 to 12.56 and we can say, "The impact of a play area on visit duration is 12.56 additional minutes *for the average customer age in our data.*" Similarly, because *PlayArea* is a binary variable, $\beta_{a1}$ is by default the slope of the line for customers without a play area. You can change that by reversing the levels of *PlayArea* (i.e., setting 1 or "Y" as the default and 0 or "N" as the change, depending on how your binary variable is formulated):

```
## R
> centered_data <- hist_data %>%
    mutate(play_area = factor(play_area, levels=c('1','0')))

## Python
centered_data_df['play_area'] = centered_data_df['play_area']
```

Changing the default level for *PlayArea* changes the coefficient for $\beta_{a1}$ from 0 to −0.07 (the slope of the upper solid line in Figure 11-24).

How should you set the defaults for your variables? It depends on the business problem at hand as well as the nature of the variable:

- In some situations, the relevant default for a numeric variable is not the average, but the minimum, the maximum, or the average in a relevant subgroup (e.g., the average age of customers with children, or the average age of customers in stores with a play area as opposed to the global average). In particular, when you're dealing with count variables such as number of children or number of calls, zero might be a better reference point than the average.

- For binary variables, the relevant default is usually the status quo. Here for example, you would set the default for *PlayArea* to 0 if you're considering adding new play areas, and to 1 if you're considering some of the existing ones.

- For categorical variables such as gender or state, unless there is a meaningful reference point, you can default to the most common category.

Setting meaningful reference points for all the variables involved in moderation has the advantage of computational simplicity and is often straightforward. However, it can get quickly cumbersome as the number of variables involved increases. Let's imagine for example that we add an interaction between state and gender to our regression:

$$VisitDuration = \beta_1 + \beta_{p1}.PlayArea + \beta_{a1}.Age + \beta_{pa1}.(PlayArea * Age) + \beta_{g1}.Gender + \beta_{s1}.State + \beta_{gs1}.(Gender * State)$$

Now $\beta_{p1}$ is the impact of a play area for customers of the reference age and gender in the reference state. "The effect of a play area is 10 for female California customers of age 43" starts to be a mouthful and can become rather meaningless if the variables are not independent: the fact that we have more female than male customers, that we have more customers in California than in any other state, and that the overall average age is 43 doesn't imply that we have a lot of 43-year-old female California customers, or any whatsoever for that matter. This brings us to the other solution, calculating the average effect in our sample.

## Calculating effects at the level of business decisions

From a scientific perspective, one of the main advantages of the previous approach is that it provides a single number for a coefficient, which can presumably be applied to completely different circumstances, or at least compared with the numbers obtained in other circumstances. However, the goal of applied analytics is not to measure things for the sake of measurement, but to guide business decisions (if you're running a data analysis and you don't know what possible decisions could come out of it, you need to have a conversation with your manager, because at least one of you is not doing their job correctly). Therefore, an alternative approach is to calculate the value of your effect variable of interest with and without that decision.

Let's imagine for example that the decision at hand for C-Mart is to pick the next store in which to set up a play area. To answer that question, we don't need to determine the single "average" effect of a play area, and trying to do so would actually be counterproductive. Instead, for each store that doesn't have a play area today, we can directly determine what would be the average additional visit duration if we added a play area. The process is as follows (the callout numbers are shared between R and Python):

```
## Python (output not shown)
def business_metric_fun(dat_df):
    model = ols("duration~play_area * (children + age)", data=dat_df) ❶
    res = model.fit(disp=0)
    action_dat_df = dat_df[dat_df.play_area == 0].copy() ❷
    action_dat_df['pred_dur0'] = res.predict(action_dat_df) ❸
    action_dat_df.play_area = 1 ❹
    action_dat_df['pred_dur1'] = res.predict(action_dat_df) ❺
    action_dat_df['pred_dur_diff'] = \ ❻
        action_dat_df.pred_dur1 - action_dat_df.pred_dur0
    action_res_df = action_dat_df.groupby(['store_id']) \ ❼
        .agg(mean_dur_diff=('pred_dur_diff', 'mean'),
             tot_dur_diff=('pred_dur_diff', 'sum'))
    return action_res_df
action_res_df = business_metric_fun(hist_data_df)
action_res_df.describe()

## R
business_metric_fun <- function(dat){
```

```
mod_model <- lm(duration~play_area * (children + age), data=dat) ❶
action_dat <- dat %>%
  filter(play_area == 0) ❷
action_dat <- action_dat %>%
  mutate(pred_dur0 = predict(mod_model, action_dat)) %>% ❸
  mutate(play_area = factor('1', levels=c('0', '1'))) ❹
action_dat <- action_dat %>%
  mutate(pred_dur1 = predict(mod_model, action_dat)) %>% ❺
  mutate(pred_dur_diff = pred_dur1 - pred_dur0) %>% ❻
  dplyr::group_by(store_id) %>% ❼
  summarise(mean_d = mean(pred_dur_diff), sum_d = sum(pred_dur_diff))
  return(action_dat)}
action_summ_dat <- business_metric_fun(hist_data)
summary(action_summ_dat)

    store_id      mean_d            sum_d
3     : 1    Min.   :10.41    Min.   :109941
4     : 1    1st Qu.:11.26    1st Qu.:129817
5     : 1    Median :11.80    Median :143079
7     : 1    Mean   :11.95    Mean   :144616
8     : 1    3rd Qu.:12.25    3rd Qu.:155481
9     : 1    Max.   :14.43    Max.   :207647
(Other):27
```

❶ Run and save the model to use it for predictions.

❷ Select the stores without a play area today.

❸ Add the predicted visit duration under the current circumstances, pred_dur0.

❹ Change the binary variable *PlayArea* from 0 to 1.

❺ Determine the predicted visit duration with the added play area, pred_dur1.

❻ Calculate the difference between the two.

❼ Aggregate at the store level, either by looking at the average or total additional duration (the average is more intuitive, but the total relates more directly to business outcomes by favoring larger stores).

We can then select the store(s) with the highest benefits of a play area. Crucially, you can check for yourself that centering numeric variables at the beginning of the process leads to the exact same final conclusions. Mathematically, that's because we're subtracting the same quantity from the two terms in a difference:

$$(VisitDuration_{i1} - mean(VisitDuration)) - (VisitDuration_{i0} - mean(VisitDuration))$$
$$= VisitDuration_{i1} - VisitDuration_{i0}$$

where *VisitDuration*$_{i1}$ and *VisitDuration*$_{i0}$ are the predicted visit duration, respectively with and without a play area (regardless of current conditions). Therefore, with a decision-centric perspective the reference points and centering, or lack thereof, are irrelevant and you don't need to worry about that anymore.

To recap: adding a moderation term changes the value and interpretation of the individual coefficients for the variables involved. This happens because the very definition of moderation is that these coefficients are not the same "everywhere," and moderation changes the baseline values at which they are measured. The individual coefficients must thus be interpreted either in relation to relevant reference points of the variables involved (which you can adjust through centering), or across our entire data set but in relation to the decision at hand.

# Conclusion

One of the key tenets of behavioral science is that "behavior is a function of the person and the environment." That phrase is generally taken to mean that we can influence behaviors by changing the environment. That is certainly true but I also like to see it as a reminder that averages are just that, averages, and it behooves the diligent behavioral analyst to dig deeper with moderation analysis. In particular, I think that some of the recent failures to replicate classic psychology experiments would be best interpreted not as meaning that "there is no effect" but that "the effect is strongly moderated by population characteristics (i.e., the person) and experimental conditions (i.e., the environment)."

This may seem to be a matter of academic interest only, but it's not. In business settings, many heated discussions where both sides have commandeered anecdotal supporting evidence can be usefully recast in terms of moderation. What is true on the East Coast may not be on the West Coast. A training program may be effective with inexperienced staff but not with experienced employees, or vice versa. By definition, a regression without moderation would not be able to shed light on either of these two cases.

This makes moderation analysis a very valuable, but often neglected, addition to the behavioral analytics tool belt. In this chapter, we saw that it can be applied both to observational and experimental data in a very simple way: add a multiplicative term between the two variables to your regression. In the next and last chapter, we'll turn to another of the key behavioral data analysis tools, mediation.

# CHAPTER 12

# Mediation and Instrumental Variables

In the last chapter, we saw that moderation allows us to open the black box of a causal relationship by revealing groups for which that relationship is stronger or weaker. *Mediation* refers to the presence of an intermediary variable between two variables in a chain; it offers a different way to probe that black box by understanding the causal mechanism at play—the "how" of the causal effect.

This has several benefits on both the causal and behavioral sides of our framework. From a causal perspective, mediation reduces the risk of false positives, and not accounting adequately for it can bias our analyses. From a behavioral perspective, mediation helps us better design and understand experiments. In a sense, mediation is nothing new, and most of the arguments in this chapter could have been summarized as "expand the chains in your CDs as much as you can, at least at the beginning." But I believe such a simplification would have been a disservice to you, because the search for mediators is at the root of many scientific discoveries. "But why?" is one of the best follow-up questions after confirming the causal relationship between two variables. Customer satisfaction increases retention, *but why*? Is it because it reduces the probability of looking for alternatives or because it increases the customer's opinion of the company?

Mediation also offers a nice stepping stone for the last tool we'll see in this book, instrumental variables (IVs). IVs, which are like mediation on steroids, allow us to answer questions that would otherwise be intractable. As promised at the beginning of the book, we'll obtain an unbiased estimate of the effect of customer satisfaction on later purchasing behavior, and IVs make that possible.

In the next section, I'll introduce you to mediation in the context of the C-Mart store play areas example from Chapter 11 and show you how mediation can make a causal-behavioral analysis more effective. Then in the second section, we'll get to the grand finale: instrumental variables.

# Mediation

Let's continue with the C-Mart example from the previous chapter and assume that C-Mart is now interested in measuring the effect of the play areas on grocery purchases (Figure 12-1).

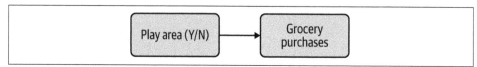

*Figure 12-1. Our relationship of interest*

Based on prior analyses, the management of C-Mart believes that visit duration is a key causal mechanism in this relationship. That is, they believe that *PlayArea* causes *VisitDuration*, which itself causes *GroceryPurchases*. Obviously, we could disregard that hypothesis and analyze the relationship displayed in Figure 12-1 directly by building the corresponding regression, CIs, etc. However, we're going to confirm and measure that causal mechanism, because taking it into account has several benefits:

- Mediation allows us to understand the mechanisms at play and generate actionable insights.

- In certain circumstances, not accounting for mediation could bias a causal estimate.

Let's review these benefits in more detail in the next two subsections, and then I'll get into the technical considerations of measuring mediation.

## Understanding Causal Mechanisms

The first benefit of identifying and measuring mediation is that it provides an explanation of the causal mechanism at hand. Correlation is still not (always) causation, but understanding what's actually happening from a behavioral perspective is a strong safeguard against spurious correlations. If you have two variables that are correlated but you're unsure whether that correlation is causal, finding and validating a mediator between them offers very strong evidence that the relationship is causal. At that point, the most likely source of error would be reverse causality—the causality flows in the opposite direction. (The alternative is that each of the three variables just happens to have a spurious correlation with the other two; that's three spurious correlations from random chance instead of one, which is a highly improbable event.)

Mediation is also a very effective complement to moderation, in both the discovery and the design phases. In the discovery phase (what I referred to as the "fishing expedition" in the previous chapter), identifying likely mediators can help the brainstorming process. Even if a mediator is not observable (e.g., a belief or emotion), simply considering it might lead you to a measurable moderator and provide a rationale for it. We saw in Chapter 11 that the relationship between *PlayArea* and *GroceryPurchases* is moderated by *Children*, something that makes immediate and intuitive sense. However, recognizing that this relationship is mediated by *VisitDuration* can give us clues to other possible moderators. For example, if the mediation is complete, the overall effect is likely to be weaker for visits near the closing time, something that would not be immediately apparent if we disregarded the role of *VisitDuration* (Figure 12-2).

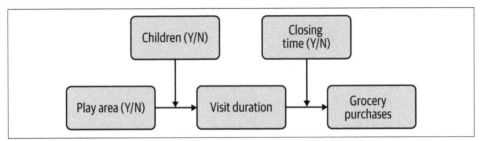

*Figure 12-2. The relationship between PlayArea and GroceryPurchases*

We could probe that hypothesis by checking whether the effect of *PlayArea* on *GroceryPurchases* is weaker for visits that start near the closing time, which would provide further information about the nature of the effect.

Mediation can also be helpful for designing or improving business processes and messages. Too often, segmentation exercises end up with "segment A shows a stronger effect than segment B." This is only useful in better targeting preexisting treatments, such as marketing campaigns. Understanding the mechanism through which the moderation acts provides us with useful insights for ideation. Having confirmed that the moderating effect of *Children* on *PlayArea* is mediated by *VisitDuration*, we could lean into it by offering a perk, like extra parking validation or a snack station near the play area. We could also replace the play areas with (potentially cheaper) mini-cinemas showing cartoons.

## Causal Biases

Mediation is not just a "nice to have" tool. In some circumstances, not accounting for it can introduce biases in our causal estimates.

The simplest case where this occurs is when we're trying to measure the (total) effect of one variable on another, but we unwittingly include a mediator in our regression as a control variable. Let's assume that we've measured the effect of having a play area on grocery purchases at the individual level, in line with the recommendations from Chapter 2. It is conceivable that a play area not only affects the visit duration of customers who would have come regardless but also draws new customers. To account for this causal path, we need to recast our CD at the store level (Figure 12-3).

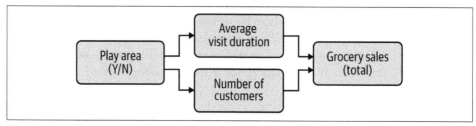

*Figure 12-3. CD recast at the store level*

Note that while average grocery purchases at the customer level are fully mediated by average visit duration, this is not the case for total grocery sales at the store level. The level at which you measure behaviors matters!

In Figure 12-3, it is clear that *CustomersNumber* is a mediator of the effect of *PlayArea* on *GrocerySales*. However, we can easily imagine that someone investigating that effect would decide to include *CustomersNumber* as a control variable in that regression: *GrocerySales* = $\beta_p$.*PlayArea* + $\beta_c$.*CustomersNumber*. After all, the size of a store's customer base certainly affects its total grocery sales, and it may have impacted the selection of stores where play areas would be built (Figure 12-4).

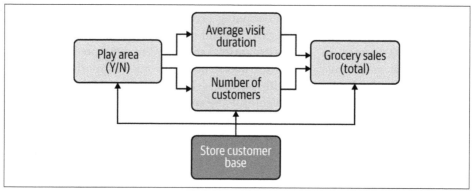

*Figure 12-4. The store's customer base is a confounder*

Figure 12-4 presents a causal conundrum: the store's customer base, measured by the number of potential customers who are within shopping distance of the store, may be a confounder of the relationship between *PlayArea* and *GrocerySales*. But at the same time, *CustomersNumber* is a mediator of that relationship. This situation calls for thoughtful controlling of the relevant regressions—that is, all the regressions where the dependent variable is *CustomersNumber* or *GrocerySales*.

This can be achieved by adding to these regressions thoughtful control variables. For instance, the number of customers one year before installing the play area is a good proxy for the customer base and by definition doesn't capture any of the effects of installing a play area. Alternatively, we could choose another proxy such as a *Rural/ Urban* categorical variable.

Once again, this illustrates the dangers of the "everything and the kitchen sink" approach to variable inclusion: just because the current number of customers is available and relevant to our situation doesn't mean that it should automatically be included as a control.

## Identifying Mediation

When I introduced the building blocks of CDs in Chapter 3, I mentioned that a *mediator* is a variable between two other variables in a chain, as in Figure 12-5.

*Figure 12-5. The effect of having a play area on grocery purchases is mediated by the duration of the visit*

*VisitDuration* is an effect of *PlayArea*, and a cause of *GroceryPurchases*. As such, it is the mediator of the effect of *PlayArea* on *GroceryPurchases* in this CD.

Assuming that the CD in Figure 12-5 gives us the whole story, *PlayArea* has no effect on *GroceryPurchases* apart from the path through *VisitDuration*; changing *PlayArea* while holding *VisitDuration* constant would not change *GroceryPurchases*. The effect of *PlayArea* on *GroceryPurchases* is said to be "completely" or "fully" mediated. Alternatively, we could imagine a situation where *PlayArea* also has a direct effect on *GroceryPurchases*, apart from the *VisitDuration* path (Figure 12-6).

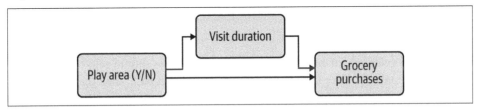

*Figure 12-6. Partial mediation*

This is called "partial" mediation. *VisitDuration* is still a mediator, even if it doesn't account for the total effect of *PlayArea* on *GroceryPurchases*. The direct effect from *PlayArea* to *GroceryPurchases* may be truly direct—there is no intermediary variable involved—or it may represent one or several other mediators we don't know of or are not interested in analyzing (in which case we have collapsed the corresponding chain or chains).

Like the search for potential moderators, the search for mediators is a trial-and-error fishing expedition, but the scope of the search (and therefore the risk of false positives) is much more limited. Since a potential mediator should be investigated only if there is a strong behavioral rationale for it, the number of candidates is considerably restricted. In addition, confirming a mediator involves several regressions, which also reduces the risk of false positives.

Given the benefits of accounting for mediators and the risks of not doing so, you should always include relevant mediators in your analysis, at least as a first step. Once you have identified which chains you're going to analyze and which you're going to disregard, you can safely collapse the latter ones (e.g., mediators between variables that play only a secondary role in your analysis).

As with moderators, the search process entails finding a potential candidate and then validating it through a Bootstrap confidence interval. In our example, *VisitDuration* is an obvious candidate, so let's see how we can confirm and measure its mediation role.

## Measuring Mediation

Measuring mediation is straightforward but a bit cumbersome. It boils down to running several regressions to estimate the following:

- The total effect of *PlayArea* on *GroceryPurchases*
- The effect of *PlayArea* on *GroceryPurchases* that is mediated by *VisitDuration*, a.k.a. the indirect effect
- The effect of *PlayArea* on *GroceryPurchases* that is not mediated by *VisitDuration*, a.k.a. the direct effect

If we find no evidence for the indirect, mediated path, we should reject our tentative mediator. Conversely, if we find no evidence for the direct path, then the effect is fully mediated. The *percentage of total effect mediated* is a common and useful way to summarize that evidence. We'll wrap up the section by considering the special case in which the mediator is a binary variable.

### Total effect

We first determine the total effect by running a regression of *GroceryPurchases* on *PlayArea* (without including *VisitDuration*):

```
## R (output not shown)
summary(lm(grocery_purchases~play_area, data=hist_data))

## Python
ols("grocery_purchases~play_area", data=hist_data_df).fit().summary()
...
                 coef    std err    t         P>|t|    [0.025   0.975]
Intercept  49.1421    0.047     1036.494    0.000    49.049   49.235
play_area  27.6200    0.079     349.485     0.000    27.465   27.775
...
```

The total effect is approximately 27.6, meaning that adding a play area increases the amount spent on groceries by $27.6 on average, *not holding the visit duration constant*.

## Mediated effect

The effect of *PlayArea* on *GroceryPurchases* mediated by *VisitDuration* can be obtained by multiplying together the effect of *PlayArea* on *VisitDuration* and the effect of *VisitDuration* on *GroceryPurchases*. This makes intuitive sense: if a play area increases the average duration of a visit by X minutes and each additional minute of visit duration increases the amount spent on groceries by $Y, then adding a play area increases the amount spent on groceries by $X * Y.

The first regression is for the arrow between *PlayArea* and *VisitDuration*. It yields a coefficient of about 12.6 (having a play area adds about 12.6 minutes to the average visit duration):

```
## R (output not shown)
summary(lm(duration~play_area, data=hist_data))

## Python
ols("duration~play_area", data=hist_data_df).fit().summary()
...
             coef  std err     t      P>|t|  [0.025     0.975]
Intercept  23.8039  0.018  1287.327  0.000  23.768   23.840
play_area  12.5570  0.031   407.397  0.000  12.497   12.617
...
```

The second regression is for the arrow between *VisitDuration* and *GroceryPurchases*. However, I will also include *PlayArea* in this regression. Looking back at Figure 12-6 and remembering the definition of a confounder, we can see that if the mediation is only partial (i.e., there is a direct arrow from *PlayArea* to *GroceryPurchases*), *PlayArea* is a confounder of the relationship between *VisitDuration* and *GroceryPurchases*. Therefore it must be included in the regression by default. Running a regression with our primary cause and our mediator as explanatory variables yields coefficients of, respectively, 0.16 (adding a play area adds about $0.16 to the average grocery purchase per visit, holding the visit duration constant) and 2.2 (adding one minute to visit duration adds about $2.20 to the average grocery purchases per visit):

```
## Python (output not shown)
ols("grocery_purchases~duration+play_area", data=hist_data_df).fit().summary()

## R
summary(lm(grocery_purchases~duration+play_area, data=hist_data))
...
Coefficients:
              Estimate Std. Error  t value Pr(>|t|)
(Intercept) -2.917728   0.047329  -61.647  < 2e-16 ***
duration     2.187025   0.001695 1290.410  < 2e-16 ***
play_area1   0.157477   0.046419    3.393 0.000693 ***
...
```

In this very simple example, these three are the only variables involved, but in real life you would have to also include any other variable with an arrow towards the dependent variable in each of your regressions.

I mentioned earlier that the primary cause should be included in the regression "by default." Sometimes, however, the primary cause and the mediator may be so closely correlated that including both of them in a regression creates multicollinearity. This is generally caused by full mediation and is evidenced by suspiciously large coefficients in opposite directions (that is, the primary cause and the mediator mostly cancel each other out) with large p-values. In the worst cases, your analysis software may even give up and return an error message without finishing the regression. Whenever including the primary cause in the regression makes the coefficient for the mediator go haywire, do not include the primary cause.

Finally, you may also encounter more complex situations, such as two mediators with an extra arrow between them, that is, one of the mediators is also a cause of the other (Figure 12-7). This is not just a theoretical possibility; it actually happens from time to time, especially with behavioral data. Situations like that one defy shortcuts and you'll need to remember what we learned in Part II: the backdoor criterion for deconfounding and other rules of CDs still apply and will let you know which variables you should and should not include in your regression.

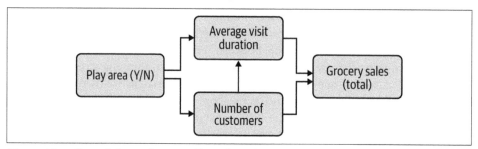

*Figure 12-7. One of the mediators affects the other mediator*

The mediated effect is equal to the product of the two coefficients along the mediation chain (i.e., the coefficient of the regression of *VisitDuration* on *PlayArea*, and the coefficient of the regression of *GroceryPurchases* on *VisitDuration*):

$$MediatedEffect \approx 12.6 * 2.2 \approx 27.5$$

From there, we can calculate the percentage of the total effect that is mediated:

$$PercentageMediated = MediatedEffect / TotalEffect \approx 27.5 / 27.6 \approx 99.5\%$$

The Bootstrap 90%-CI for this percentage is approximately [0.9933; 0.9975].

## Direct effect

The direct, unmediated effect is equal to the coefficient for *PlayArea* in the regression of *GroceryPurchases* on *PlayArea* and *VisitDuration*: $UnmediatedEffect \approx 0.16$. From there, we can calculate the percentage of the total effect that is unmediated:

$$PercentageUnmediated = UnmediatedEffect / TotalEffect \approx 0.16 / 27.6 \approx 0.5\%$$

In other words, the total effect is technically not fully mediated. For practical purposes, however, we can neglect the unmediated effect. Note that "unmediated effect" is always expressed in relation to a specific mediator. If you had two mediators that together completely mediated a total effect, the mediated effect of the first one would be equal to the unmediated effect of the second one and vice versa.

In situations where you encounter multicollinearity and are unable to include the primary cause in the regression as described earlier, the effect is most likely fully mediated. You can and should check that this is the case by calculating the percentage of the total effect that is unmediated as the residual of the mediated effect:

$$PercentageUnmediated = (TotalEffect - MediatedEffect) / TotalEffect$$

If you obtain an unmediated effect that is economically significant through that approach, this means you most likely have a more complex causal structure on your hands than you thought. Then it's time to revisit your CD with a critical eye. Maybe your primary cause and your mediator share a further joint cause? Or could there be several mediators with interrelations between them?

The mediator in this example was positively correlated with both its cause and its effect. Mediators can also have a negative effect if exactly one of the two coefficients along the chain is negative. If that happens, and there is another mediator or a direct effect between our cause and effect of interest, then the mediator will *reduce* the total effect. In that case, your first mediator could represent, say, −25% of the total effect while the direct effect (or another mediator) would represent 125% of the total effect, but the sum of the proportions of effects will still be 100%. This is perfectly normal and expected, so don't let it trip you up.

If you find that confusing, you're in good company. Social scientists debated for decades whether or not having a significant total effect should be a prerequisite to mediation analysis. But there are a variety of perfectly legitimate regulating or self-regulating phenomena where a mediator compensates for the direct effect. For example, C-Mart may find that its price changes do not affect its sales volume as expected because they cause corresponding price changes from its competitors.

### When the mediator is a binary variable

The mediator in this example was a numeric variable, so we could easily obtain the mediated effect by multiplying the coefficients for the two arrows involved. When the mediator is a binary variable, it is still possible to quantify mediation analytically, i.e., by using equations, but the formulas get more convoluted.

If we call the cause of interest $X$, the mediator $M$, and the effect of interest $Y$, the regression equations for the mediator and the final effect become:

$$P(M = 1) = \text{logistic}(\alpha_0 + \alpha_X.X)$$

$$Y = \beta_0 + \beta_X.X + \beta_M.M$$

Note that the first equation is now a logistic regression, as appropriate for a binary variable. Instead of predicting the value of M as with a linear regression, we now predict the probability that it will take the value 1, $P(M = 1)$. We can substitute that probability in the second equation:

$$Y = \beta_0 + \beta_X.X + \beta_M.P(M = 1)$$

The direct effect is still straightforward to calculate: if X increases by 1, the direct effect increases Y by $\beta_X$. But the indirect effect now presents an additional challenge because the impact of X on M is not linear. Therefore, the impact of X on M has to be determined *for a certain value of X*. This issue is similar to the one we encountered with moderation in Chapter 11, and the possible solutions are the same:

- Define a global reference point, such as the average value of X in our data.
- Or calculate the mediated effect and percentage mediated for each row in our data and then calculate their respective averages.

As was the case in Chapter 11, I recommend the second approach, modified as needed to fit the business decision at hand.

# Instrumental Variables

Mediation represents a great addition to the behavioral data analysis toolbox in itself, but it's also a stepping stone to another powerful tool called instrumental variables (IVs). In a nutshell, IVs leverage a known mediation relationship to reduce confounding biases in our coefficients.

One of the most powerful use cases is to use an experiment to answer a broader, and often harder, question. I'll illustrate this application with an example involving customer satisfaction, one of the most watched business metrics, but also admittedly one of the hardest to measure.

As first mentioned in Chapter 2, AirCnC's leadership wants to know the impact of customer satisfaction (*CSAT*) on one of their key performance indicators, the amount spent during the six months following a booking (*M6Spend*). We'll reuse the data from the experiment in Chapter 10 in which we explored the impact on customer satisfaction of a change in call center procedures: namely, "Instead of apologizing repeatedly when something went wrong, the call center reps should apologize at the beginning of the interaction, then get into 'problem-solving mode,' then end up offering several options to the customer."

## Data

The GitHub folder for this chapter (*https://oreil.ly/BehavioralDataAnalysisCh12*) contains a copy of the experimental data from Chapter 10. This time we'll include the M6Spend variable in our analysis. Table 12-1 outlines the variables in our data for this chapter.

*Table 12-1. Variables in our data*

|  | Variable description | chap10-experimental_data.csv |
|---|---|---|
| *Center_ID* | Categorical variable for the 10 call centers | ✓ |
| *Rep_ID* | Categorical variable for the 193 call center reps | ✓ |
| *Age* | Age of customer calling, 20-60 | ✓ |

| | Variable description | chap10-experimental_data.csv |
|---|---|:---:|
| *Reason* | Reason for call, "payment"/"property" | ✓ |
| *Call_CSAT* | Customer satisfaction with call, 0-10 | ✓ |
| *Group* | Experimental assignment, "ctrl"/"treat" | ✓ |
| *M6Spend* | Amount spent on bookings within 6 months of a booking | ✓ |

## Packages

In this section, we'll use the following specific packages for instrumental variables:

```
## Python
from linearmodels.iv import IV2SLS
```

```
## R
library(ivreg)
```

## Understanding and Applying IVs

Once you're familiar with CDs and mediation, the idea behind IVs becomes reasonably simple to express:

> Let's assume you have a fully mediated relationship between two variables, and the relationship between the mediator and the final variable is confounded. Then you can obtain an unbiased estimate for that relationship by dividing the coefficient for the total effect by the coefficient for the first arm of the mediation (i.e., the relationship between the first variable and the mediator).

---

### The History of IVs, a.k.a. the Roots of Causal Diagrams

Instrumental variables were invented by Philip and Sewall Wright, a father and son who were also among the inventors of path analysis, the precursor of causal diagrams. Sadly, that origin story was lost for decades, and IVs were used as a stand-alone tool by economists. For example, the authors of leading introductory books on econometrics, Angrist and Pischke (2009, 2014), present IVs without ever introducing causal diagrams (although they do mention the Wrights).

In my experience, using instrumental variables without CDs is a head-scratching, wall-staring exercise. You need to postulate a potential instrument seemingly out of thin air, and then justify the necessary assumptions through lengthy discussions. It can certainly be done, but it requires an implicit knowledge of the business or economic situation that comes only from experience and is hard to communicate. Conversely, as we'll see here, if that knowledge is expressed in the form of a carefully crafted and iterated-upon CD, applying IVs is a matter of course.

---

To see what this looks like in our example, let's start by drawing the CD of our variables of interest. We want to measure the causal relationship between *CSAT* and *M6Spend*. It's plausible that a high CSAT increases the amount spent on bookings in the following months, but the relationship is also subject to unmeasured confounders, including personality traits such as openness. Finally, we have data on our experimental treatment, which we know affects *CSAT* thanks to our experiment in Chapter 10 (Figure 12-8).

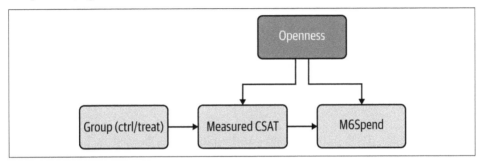

*Figure 12-8. The CD for our variables of interest*

As you can see, in this CD, *CSAT* is a mediator between *Group* and *M6Spend*, but the relationship between *CSAT* and *M6Spend* is biased upward by the confounding effect of *Openness*.

In an ideal world, we would have data about the *Openness* variable and we would be able to run the two true regressions (Equation 12-1 and Equation 12-2).

*Equation 12-1.*

$$CSAT = \beta_{g1}.Group + \beta_{o1}.Openness$$

*Equation 12-2.*

$$M6Spend = \beta_{c2}.CSAT + \beta_{o2}.Openness$$

But even if we got data about Openness specifically (e.g., through a survey), how could we be sure that there isn't another confounder lurking around the corner?

Taking a step back from the math for a second, this question takes us to the very heart of behavioral analytics: customer satisfaction is an essential criterion for success in business, but because it is so strongly influenced by a myriad of unobserved individual characteristics that impact individual behaviors, we can't hope to identify and control for all of these characteristics. This issue can't be resolved satisfactorily outside of a causal-behavioral framework, but it is straightforward to handle within it.

Back to the math—let's go through the intuition laid out earlier:

1. Calculate the coefficient for the leftmost relationship, between *Group* and *CSAT*.
2. Calculate the coefficient for the total effect of *Group* on *M6Spend*.
3. Calculate the coefficient for the effect of *CSAT* on *M6Spend* by dividing the total effect from step 2 by the coefficient for the leftmost relationship from step 1.

## Step 1: Leftmost relationship

Because *Group* is a random assignment, uncorrelated with *Openness*, we can run the following regression instead of Equation 12-1:

$$CSAT = \beta_{g1}.Group$$

Our estimate for $\beta_{g1}$ is unbiased and can be plugged as needed in Equation 12-1, because it is the true causal coefficient.

## Step 2: Total effect

The equation for the total effect (Equation 12-3) is called the *reduced* regression (and we'll index it by "r") because it collapses the chain between the variables.

*Equation 12-3. Equation R*

$$M6Spend = \beta_{gr}.Group \text{ (Eq. R)}$$

For the same reasons as for the first arm of the mediation, our estimate for $\beta_{gr}$ is unbiased.

## Step 3: Relationship of interest

This is where the magic happens: as discussed in the previous section about mediation, $\beta_{gr} = \beta_{c2} * \beta_{g1}$. We can rewrite this equation as $\beta_{c2} = \beta_{gr} / \beta_{g1}$. Because all variables right of the equal sign are unbiased, the one left of it is as well. This estimate for $\beta_{c2}$ is unbiased.

In other words, we can deconfound the relationship between two variables if we can find a variable (called the *instrument*) that is a cause of our cause of interest but is otherwise unrelated to both the confounder and the effect of interest:

- The first condition (called the *independence assumption*) is necessary for the reduced regression to be unbiased, but fortunately it is always true with a random assignment.
- The second condition (called the *exclusion restriction*) can be rephrased as saying that the relationship between the instrument and the effect of interest must be fully mediated by the cause of interest. It is necessary for the equation $\beta_{gr} = \beta_{c2} * \beta_{g1}$ to be true. Unfortunately, it can't be proven mathematically (that would require knowing the coefficient for the second arm of the mediation, which is what we're looking for!) and must be assumed based on qualitative causal considerations—in this case, for example, that the experimental group assignment is unlikely to affect *M6Spend* outside of the *CSAT* chain.

## Measurement

That was the intuition. We could certainly calculate all the corresponding regressions by hand, but as we spoiled twenty-first-century data analysts have come to expect, there is a package for that.

First, we'll do two sanity checks by running the linear regressions for the first arm of the mediation and for the total effect (i.e., the reduced equation). If either of these yields a coefficient very close to zero (as determined by the Bootstrap CI), that would jeopardize our IV regression. In most cases, you'll want to include some other covariates that are causally related to your variables of interest. In our example from Chapter 10, *Age* (the age of the caller) and *Reason* (the reason for the call) were predictive of *CSAT* and included beside *Group*. We should include them here as well:

```
## Python (output not shown)
ols("call_CSAT~group+age+reason", data=exp_data_df).fit(disp=0).summary()
ols("M6Spend~group+age+reason", data=exp_data_df).fit(disp=0).summary()

## R
summary(lm(call_CSAT~group+age+reason, data=exp_data))
summary(lm(M6Spend~group+age+reason, data=exp_data))
..
Coefficients:
               Estimate Std. Error t value Pr(>|t|)
(Intercept)    4.103826   0.011790  348.07  <2e-16 ***
grouptreat     0.540633   0.006291   85.94  <2e-16 ***
age            0.020202   0.000280   72.14  <2e-16 ***
reasonproperty 0.200590   0.006600   30.39  <2e-16 ***
...
Coefficients:
            Estimate Std. Error  t value         Pr(>|t|)
(Intercept) 99.93195   0.43976  227.242         < 2e-16 ***
grouptreat   1.61687   0.23465    6.891 0.00000000000557 ***
age         -1.46785   0.01044 -140.536         < 2e-16 ***
```

```
reasonproperty   0.44458     0.24615     1.806              0.0709 .
...
```

Fortunately, both coefficients are safely distinct from zero, so we can move to our IV regression.

### Python code

In Python, we'll use the package `linearmodels`:

```
## Python
iv_mod = IV2SLS.from_formula('M6Spend ~ 1 + age + reason + [call_CSAT ~ group]',
                             exp_data_df).fit()
iv_mod.params
Out[8]:
Intercept           87.658610
reason[T.property]  -0.155326
age                 -1.528264
call_CSAT            2.990706
Name: parameter, dtype: float64
```

The syntax of the `IV2SLS.from_formula()` function is almost the same as for `ols()`. The predicted variable, our effect of interest, goes to the left of the tilde sign ("~") and the predictors go to the right of it, with the first-stage regression written between brackets. Two things to note:

- You need to explicitly include a constant ("1") among the predictors.
- The other covariates related to your variables of interest (here, *Age* and *Reason*) should be included here as well, outside of the brackets. Note that they will automatically be included in the first-stage regression. Your formula for the first-stage regression only needs to include the mediator, a.k.a. our cause of interest, and the instrument.

The output of the Python function is limited, but that's all we need. The effect of *call_CSAT* on *M6Spend* is about $2.99 per unit, about $1 less than the naive, biased regression of *M6Spend* on *call_CSAT* would suggest:

```
## Python
ols("M6Spend~call_CSAT+age+reason", data=exp_data_df).fit(disp=0).summary()
...
                    coef    std err    t      P>|t| [0.025  0.975]
Intercept          83.2283  0.536    155.302  0.000  82.178  84.279
reason[T.property] -0.3582  0.245     -1.461  0.144  -0.839   0.122
call_CSAT           4.0019  0.076     52.767  0.000   3.853   4.151
age                -1.5488  0.010   -147.549  0.000  -1.569  -1.528
...
```

The Bootstrap 90%-CI for the unbiased effect is approximately [2.26; 3.89].

## R code

In R, we'll use the package `ivreg`:

```
## R
> iv_mod <- ivreg::ivreg(M6Spend~call_CSAT + age + reason | group + age + reason,
    data=exp_data)
> summary(iv_mod)

Call:
ivreg::ivreg(formula = M6Spend ~ call_CSAT + age + reason | group +
    age + reason, data = exp_data)

Residuals:
   Min    1Q Median    3Q    Max
-86.82 -35.01 -17.94  19.92 706.58

Coefficients:
               Estimate Std. Error  t value      Pr(>|t|)
(Intercept)    87.65861    1.93745   45.244       < 2e-16 ***
call_CSAT       2.99071    0.43165    6.929 0.00000000000426 ***
age            -1.52826    0.01358 -112.540       < 2e-16 ***
reasonproperty -0.15533    0.25968   -0.598          0.55

Diagnostic tests:
                 df1    df2 statistic p-value
Weak instruments   1 231655  7384.847   <2e-16 ***
Wu-Hausman         1 231654     5.667   0.0173 *
Sargan             0     NA        NA      NA
---
Signif. codes:  0 '***' 0.001 '**' 0.01 '*' 0.05 '.' 0.1 ' ' 1

Residual standard error: 56.12 on 231655 degrees of freedom
Multiple R-Squared: 0.0925,     Adjusted R-squared: 0.09249
Wald test:  7019 on 3 and 231655 DF,  p-value: < 2.2e-16
```

Thankfully, the authors of the `ivreg` package have made a conscious effort to make the `ivreg()` function as similar as possible to `lm()`, in both its syntax and its output. The only difference in the formula is that there are now two lists of regressors, separated by a vertical bar "|". Here's where the variables need to go:

- The cause of interest, here *call_CSAT*, appears only on the left of the bar.
- The instrument(s), here *Group*, appear only on the right of the bar.
- The other explanatory variables that contribute to our cause or effect of interest, here *Age* and *Reason*, appear on both sides of the bar.

The output of ivreg() is also very similar to the output of lm(). Our value of interest is the coefficient for *call_CSAT*. ivreg() also returns the results of several diagnostics for IV, which test the strength of various relationships in our model. Here again, the effect of *call_CSAT* on *M6Spend* is about $2.99 per unit. The corresponding Bootstrap 90%-CI is approximately [2.26; 3.89].

 In our simple example, this difference between the two estimates comes from *Openness*. In real life, it's unlikely that we would be able to confidently identify all the confounders at play, which we could instead, if you will, label "unknown psychological stuff." However, even without measuring these confounders, we can measure their impact on *M6Spend* through *CSAT*: a change in these confounders that causes *CSAT* to increase by 1 point also causes *M6Spend* to increase by about $1. Now we can go run surveys to measure openness or any other unknown psychological stuff with the knowledge that there is, so to speak, $1 worth of confounding on *M6Spend* per point of *CSAT* to account for. We know what we don't know, which, if you ask me, is pretty cool.

## Applying IVs: Frequently Asked Questions

In the preceding example, we've leveraged IVs for the downstream analysis of experimental data: using experimental data to deconfound causal relationships. This is one of their most straightforward and potent uses, but it's not the only one. Once you get familiar with IVs, you'll start asking yourself "what about...?" Building complete examples for each potential use case would be redundant, but it's worth calling out the most common ones:

*Can I use IVs with purely observational data?*
  Yes, and the process is exactly the same as with experimental data, but because the independence assumption is not a given in that case, you'll need to make sure that it holds. I have often come up with a potential instrument only to realize a bit later that there was another connection between the instrument and the final effect lurking in the background.

*Can I use IVs with a binary final effect?*
  In such a case, the relationships between coefficients are significantly more complicated than they are with linear regressions all along. The R package ivpro bit() allows you to run a probit regression with an instrument, but as far as I know, there is no such solution for logistic regression.

# Conclusion

Here we are. At the beginning of the book I promised that we would measure the causal impact of customer satisfaction on a business metric, and we did just that: "An increase of stated customer satisfaction by one unit increases spending in the following six months by $2.99." No lengthy caveat and footnotes, no "correlation is not causation" hand-waving—this is as clear-cut a result as it gets. And hopefully, a whole world of opportunities is opening in your mind. Customer satisfaction, loyalty program memeberships, brand perception: measuring the business implications of all these nebulous and biased concepts is within your reach. And chances are, you already have the data you need. Did someone in marketing run an experiment two years ago offering a discount to customers if they signed up for your loyalty program? Then it's simply a matter of pulling the corresponding data and applying the one-line formula for IV regression. Of course, that simplicity isn't coming out of nowhere. It's been a long time in the making, as it requires:

- Well-defined and understood variables for customer satisfaction and spending, as we saw in Part I
- The right causal diagram, as we discovered in Part II
- Tools that allow us to handle uncertainty without having to memorize a bunch of statistical tests, as we saw in Part III
- Well-designed and well-analyzed experiments, as we explored in Part IV
- Finally, an understanding of moderation and mediation, as we learned in Part V

To end on a slightly less formal note: it is often noted that children appear to have an infinite curiosity for the world ("Why is the sky blue?"). Of course, that curiosity is not actually infinite, and at some point most children stop asking so many questions. It is my sincere hope that this book will have rekindled some of that childlike curiosity in you—that you'll let yourself be intrigued by the world (especially the people) around you and you'll wonder, "But why?" Of course, before claiming any credit for that, I'll have to contemplate the very real possibility that I got the causality wrong and that you had it all in you from the start (Figure 12-9).

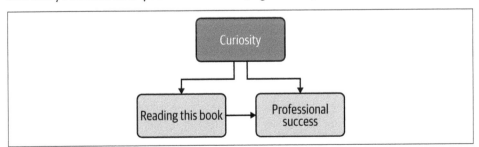

*Figure 12-9. Did I mention that correlation is not causation?*

# Bibliography

Aberson, Christopher L. *Applied Power Analysis for the Behavioral Sciences*. Abingdon, UK: Routledge, 2019.

Angrist, Joshua D., and Jörn-Steffen Pischke. *Mostly Harmless Econometrics: An Empiricist's Companion*. Princeton, NJ: Princeton University Press, 2009. A go-to reference for applied econometrics, with graduate-level math and statistics.

Angrist, Joshua D., and Jörn-Steffen Pischke. *Mastering "Metrics": The Path from Cause to Effect*. Princeton, NJ: Princeton University Press, 2014. A more accessible version of their 2009 classic, with added *Kung Fu Panda* references!

Antonio, Nuno, Ana de Almeida, and Luis Nunes. "Hotel booking demand datasets." *Data in Brief* 22 (Feb. 2019): 41-49. *https://doi.org/10.1016/j.dib.2018.11.126*.

Bertrand, Marianne, and Sendhil Mullainathan. "Are Emily and Greg more employable than Lakisha and Jamal? A field experiment on labor market discrimination." *American Economic Review* 94, no. 4 (2004): 991-1013.

Cohen, Jacob. *Statistical Power Analysis for the Behavioral Sciences*, 2nd Edition. Abingdon, UK: Routledge, 2013. A classic on power analysis. Like most classics, it is somewhat dated on computer implementation, but still a great resource to think more deeply about what statistical power is.

Cunningham, Scott. *Causal Inference: The Mixtape*. New Haven, CT: Yale University Press, 2021. A very accessible variation on the Angrist/Pischke books, still targeted primarily to academics.

Davison, A. C., and D. V. Hinkley. *Bootstrap Methods and Their Application*. Cambridge, UK: Cambridge University Press, 1997.

Efron, Bradley, and R. J. Tibshirani. *An Introduction to the Bootstrap*. Abingdon, UK: Chapman and Hall/CRC, 1994. Efron is the "inventor" of the Bootstrap, and this book provides a good introduction to it, for people with at least an undergrad understanding of math and statistics.

Eyal, Nir. *Hooked: How to Build Habit-Forming Products*. New York: Portfolio, 201.

Funder, David C. *The Personality Puzzle*. New York: W. W. Norton & Company, 2016. A great "advanced introduction" to personality psychology that bridges the gap between pop science and academic research.

Gelman, Andrew, and Jennifer Hill. *Data Analysis Using Regression and Multilevel/ Hierarchical Model*. Cambridge, UK: Cambridge University Press, 2006. Provides a deeper analysis of multilevel (i.e., clustered) data, at the cost of a higher threshold for mathematical and statistical knowledge.

Gerber, Alan S., and Donald P. Green. *Field Experiments: Design, Analysis and Interpretation*. New York: W. W. Norton & Company, 2012. My go-to book for a deeper perspective on experimental design. It contains a lot of statistical and mathematical formulas but these should be understandable by anyone with even a moderate quantitative background. My recommended resource if you want to analyze experimental data with statistical tests.

Glennerster, Rachel, and Kudzai Takavarasha. *Running Randomized Evaluations: A Practical Guide*. Princeton, NJ: Princeton University Press, 2013. An accessible introduction to experimentation in the context of development economics and aid. Contains extensive discussions of the empirical challenges of running experiments in the real world (i.e., not online).

Gordon, Brett R., et al. "A Comparison of Approaches to Advertising Measurement: Evidence from Big Field Experiments at Facebook." *Marketing Science*, INFORMS, 38, no. 2 (2019): 193-225.

Hayes, Andrew F. *Introduction to Mediation, Moderation, and Conditional Process Analysis: A Regression-Based Approach*, 2nd Edition. New York: Guilford Press, 2017.

Hernan, Miguel. *Causal Diagrams: Draw Your Assumptions Before Your Conclusions*. edX.org online course (*https://oreil.ly/dPtnt*) (accessed May 6, 2021).

Hubbard, Douglas W. *How to Measure Anything: Finding the Value of Intangibles in Business*. Hoboken, NJ: Wiley, 2010. A very penetrating and accessible perspective on measurement in business.

Jose, Paul E. *Doing Statistical Mediation & Moderation*. New York: Guilford Press, 2013. A useful complement to Hayes (2017), even though it's more focused on academic research and uses more arcane software (that is, arcane outside of academia and related fields!).

Josse, Julie, Nicholas Tierney, and Nathalie Vialaneix. "CRAN Task View: Missing Data." The Comprehensive R Archive Network, *https://cran.r-project.org/web/ views/MissingData.html*.

Kahneman, Daniel. *Thinking, Fast and Slow*. New York: Farrar, Straus and Giroux, 2013. A behavioral science classic by one of the big names in the field.

Little, Roderick J. A., and Donald B. Rubin. *Statistical Analysis with Missing Data*, 3rd Edition. Hoboken, NJ: Wiley, 2019. The third edition of the classic book on missing data analysis, written by prominent researchers in the field. Useful for a review of statistical theory.

---

Meadows, Donella H. *Thinking in Systems: A Primer*. White River Junction, VT: Chelsea Green Publishing, 2008. A great general introduction to system thinking.

Pearl, Judea. *Causality*. Cambridge, UK: Cambridge University Press, 2009. Pearl's earlier book on causality, with detailed graduate-level math.

Pearl, Judea, and Dana Mackenzie. *The Book of Why: The New Science of Cause and Effect*. New York: Basic Books, 2018. The most approachable introduction to causal analysis and causal diagrams I have encountered so far, by one of the prominent researchers in the field.

Senge, Peter M. *The Fifth Discipline: The Art & Practice of the Learning Organization*. New York: Currency, 2010.

Shipley, Bill. *Cause and Correlation in Biology: A User's Guide to Path Analysis, Structural Equations and Causal Inference with R*, 2nd Edition. Cambridge, UK: Cambridge University Press, 2016. You're not a biologist? Neither am I. This book has still helped deepen my understanding of causal diagrams, and with the limited number of books on the topic, beggars can't be choosers.

Taylor, Aaron B., and David P. MacKinnon. "Four applications of permutation methods to testing a single-mediator model." *Behavioral Research Methods* 44, no. 3 (Sep. 2012): 806-44.

Thaler, Richard H., and Cass R. Sunstein. *Nudge: Improving Decisions About Health, Wealth, and Happiness*. New York: Penguin, 2009.

van Buuren, Stef. *Flexible Imputation of Missing Data*, 2nd Edition. Abingdon, UK: Chapman and Hall/CRC, 2018. A very accessible presentation by the author of the mice package for R, with a lot of concrete examples. The author made the book freely available on his personal page, *https://stefvanbuuren.name/fimd*.

VanderWeele, Tyler. *Explanation in Causal Inference: Methods for Mediation and Interaction*. Oxford, UK: Oxford University Press, 2015. An excellent bridge between the causal analysis literature and the social sciences' literature on mediation and moderation. Covers a significant amount of more advanced material not found in the other references.

Wendel, Stephen. *Designing for Behavior Change: Applying Psychology and Behavioral Economics*, 2nd Edition. Sebastopol: O'Reilly, 2020.

Wilcox, Rand R. *Fundamentals of Modern Statistical Methods: Substantially Improving Power and Accuracy*, 2nd Edition. New York: Springer, 2010. We routinely assume that our data is normally distributed "enough." Wilcox shows that this is unwarranted and can severely bias analyses. A very readable book on an advanced topic.

# Index

## A

A/B tests
  about causal analytics, 5
  complier average causal effect (CACE) for
    encouragement design, 227
  powerful but narrow, 178
  (see also experimental design)
achieved significance level (ASL), 160
actions
  about, 25
  causal diagram variable selection, 69
  data and ethical considerations, 26
  intention-action gap, 24
  modeling human behavior, 25
  uplift analysis, 264
age as behavioral determinant, 22
aggregate metrics, 32
  linear algebra and aggregating variables, 56
  transformations of causal diagrams
    aggregating variables, 55
    slicing or disaggregating variables, 54
AirCnC (Air Coach and Couch)
  call center problem-solving mode
    about business problem, 233
    analyzing the experiment, 253
    data and packages for example, 236
    hierarchical modeling, 235, 237-242
    planning the experiment, 234-236
    power analysis, 245-252
    power curve, 249
  connecting data and behavior, 28, 30, 31, 35
  customer satisfaction and purchasing, 307
  missing data, 107
    (see also missing data)

1-click booking, 173
  (see also random assignment)
stratified randomization
  about, 203, 215
  about business problem, 203
  analyzing experimental results, 223-231
  customer satisfaction negative impact,
    205-207
  data and packages for example, 208
  planning the experiment, 205-208
  power analysis with Bootstrap, 216-223
  random assignment, 209-216
alternative hypothesis, 187, 188, 193
always-takers, 229
Anaconda Spyder, xv
analytics, types of, 4
ancestor in causal diagrams, 47
Angrist, Joshua D., 308
as.binary() function, 249
  package, 236

## B

B.E.A.N. (beta, effect size, alpha, sample size
  N), 188
backdoor criterion (BC), 97-103
behavior as action, 25
  (see also actions; business behaviors; human
    behavior)
behavior blocker granularity, 25
behavior modification reference books, 26
behavioral levels, 184
behavioral logic
  call center problem-solving mode, 236
  moderator plus, 271

random assignment, 179-181
stratified randomization, 208
behavioral variable characteristics, 32
Berkson paradox, 16
beta (β) as statistical significance, 188
biases in data and analyses
aggregate metrics, 32
Berkson paradox causing, 17
intuition subject to biases, 41
mediation
causal biases, 299
mediation unaccounted for, 297
missing data handling introducing, 107,
126, 137
What You See Is All There Is, 67
binary variables
binary dependent variable regression, 43
Cramer's V for categorical variables, 81
mediator as, 306
missing data safe to drop, 114
numeric with 0/1 values, 78
one-hot encoding categorical variables, 213
bivariate graphs to visualize missing data, 110
block() function, 214
package, 209, 236
blocked paths, 99
book coding conventions, xiv
supplemental material online, xvii
The Book of Why (Pearl and Mackenzie), 11,
16, 101
booking cancellation and deposit type CD,
64-89
Bootstrap
about package required, xiv
about uncertainty, 147
confidence intervals, 150
about statistics, 151
code, xvi, 154, 155
introduction to, 148-157
moderation analysis validation, 287-289
number of samples, 153, 164
optimizing in Python, 169
optimizing in R, 167
p-value simulation, 159
power analysis, 192-199
stratified randomization, 216-223
regression analysis code, 157-160
small sample of proportions, 155-157
small sample size

mean of simulated samples, 151
number of samples, 153
small sample with outlier, 148
about Bootstrap, 147, 150
about business problem, 148
number of samples, 164
smoothing with larger sample size, 153
when to use Bootstrap, 160-164
business behaviors
causal diagram building, 75
customer engagement, 31
data and ethical considerations, 27
modeling human behavior, 26
business goals
booking profit versus CX, 205-207
call center problem-solving mode, 234
measuring CSAT effect on future purchasing, 307
1-click booking site, 175
theory of change, 174

C
C-Mart fictional supermarket chain
Bootstrap for regression analysis, 157-160
Bootstrap for uncertainty, 155
about uncertainty, 147
cake time study, 148-154
introduction to Bootstrap, 148-157
business behavior data collection, 27
causal diagrams, 40
chains, 46
collapsing chains, 48
expanding chains, 49
experience and baking time, 157
feedback loops, 58
forks, 50-52
mediator values, 48
paths, 61
substitution effects, 57
deconfounding with causal diagrams
about, 92
backdoor criterion, 97-103
disjunctive cause criterion, 94-97
science of, 102
mediation
about, 297
business problem, 298
causal biases, 299
data for example, 307

instrumental variable measurement, 311-314
moderation complement, 298
moderation analysis
average effect redistributed, 276-278
calculating effects at decision level, 293
data and packages, 258
effect size calculations, 277
interacting moderators, 283-286
interactions, 265
interpreting individual coefficients, 289-295
moderated moderation, 283-286
parallel multiple moderators, 281-283
segmenting observational data, 259
self-moderation alerting to hidden moderator, 280
setting meaningful reference points, 291
validating with Bootstrap, 287-289
when to apply moderation, 275
personal characteristics of customers, 21
variable selection, 8
adding variables, 11
adding wrong variables, 14
confounder in action, 9
categorical variables
business-centric, 67
causal diagram variables, 77
Cramer's V for, 81
hierarchical linear modeling for, 237-242
missing data safe to drop, 114
multiple imputation in Python, 139
one-hot encoding, 213
relationships between, 81
relationships between numeric and, 84
moderating the effect, 274
causal analytics
about, 5
human complexity, 5
about analytics, 4
coefficient of correlation, 8
predictive analytics versus, 4
regression demonstrating, 8
causal diagrams (CDs)
about, 39, 63
about causal-behavioral framework, 3, 39
about goal of, 63
behaviors represented, 41
breaking down into blocks, 94

building from scratch, 89
simplifying, 88
actions, 69
base cancellation rate by deposit type, 65
business behaviors, 75
business problem, 65
cognition and emotions, 71
confounder, 66
data set, 64
expanding iteratively, 86
identifying further causes, 87
intentions, 70
iterating, 88
personal characteristics, 72
process overview, 64
proxies for unobserved variables, 86
simplest possible, 63
time trends, 75
variables to include, 67-76
variables validated via data, 77-85
collaboration tool, 67
data represented, 42-46
correlation without direct causality, 50
error terms, 45
probabilistic graphical models, 46
deconfounding with
about business problem, 92
backdoor criterion, 97-103
disjunctive cause criterion, 94-97
science of, 102
definition, 40
direct relationships, 47, 50
family tree structure, 47
first CD C-Mart iced coffee sales, 40
fundamental structures of
about, 46, 52
chains, 46-50
clear rectangles as observed variables, 40
collapsing and expanding not indicated, 50
collapsing chains, 48
collapsing chains around forks, 51
colliders, 52
direction of arrow as causality, 57
expanding chains, 49
forks, 50-52
M-pattern, 101
shaded rectangles or ovals as unobserved variables, 40, 121

unknown forks, 51
indirect relationships, 47, 50
instrumental variable roots, 308-311
intuitive understanding of causality, 41
linear regressions, 44, 45
logistic regressions, 43
mediators, 48
missing data cause diagnosed, 121-124
    adding missingness variable, 123
parent/child relationships, 46
subjectivity introduced by, 41
transformations of, 53
    aggregating variables, 55
    cycles, 57
    managing cycles, 58
    paths, 60
    slicing or disaggregating variables, 54
causal paths, 99
causal-behavioral framework
    about, xii, xiii, xvii
    about goal of, 3
    mediation bridging, 297
causes
    contributing factor in probabilistic sense, 22
    correlation implying causation, 39, 78, 84
        causal mechanisms explained, 298
        coefficient of correlation as measure of
            causal effect, 8
        identifying moderators in historical data,
            272
        mediation, 298
        white lies, 85
    intuitive understanding of causality, 41
    leading indicators as, 175
    mediation, 297
        causal mechanisms explained, 298
    missingness and causality interaction
        unknown, 126
    primary causes
        demographics as, 21, 108
        exogenous variables as, 108
    quartiles for numeric causes, 274
CDs (see causal diagrams)
central estimate, 153
    Bootstrap number of samples, 153
    when to use Bootstrap, 160-164
chains in causal diagrams, 46-50
    about, 46, 52
    causal paths, 99

collapsing and expanding not indicated, 50
collapsing chains, 48
    around forks, 51
    transitivity property, 49
expanding chains, 49
mediation as expanding chains, 297
    (see also mediation)
paths, 61
slicing or disaggregating variables, 54
child in causal diagrams, 47
cluster randomization, 236
code
    about functional-style programming, xvi
    about packages and parameters, xiv
    book coding conventions, xiv
    supplemental material online, xvii
coefficient of correlation as measure of causal
    effect, 8
cognition in human behavior
    causal diagram building, 71
    data and ethical considerations, 23
    measuring effect on behavior, 31
    model of behavior, 23
colliders in causal diagrams, 52
    M-pattern, 101
    noncausal paths, 99
    paths, 61
    slicing or disaggregating variables, 54
combinations via factorials, 246
complier average causal effect (CACE) for
    encouragement design, 226-231
complier average causal effect (CACE) for
    mandatory intervention
    formula, 229
confidence intervals (CI)
    Bootstrap
        about, 150
        about statistics, 151
        code, xvi, 154, 155
        number of samples, 153
        when to use Bootstrap, 160-164
    calculating via regression, 149
        standard error relationship, 149
    central estimate, 153
        Bootstrap number of samples, 153
        when to use Bootstrap, 160-164
    coverage, 192
    p-value simulation with Bootstrap, 159
    power/significance trade-off, 222

confounders
    about, 91
    adding variables, 11
        multicollinearity, 13
    aggregate metrics, 32
    C-Mart fictional supermarket data, 9
    collapsing chains around forks, 51
    deconfounding with causal diagrams
        about business problem, 92
        backdoor criterion, 97-103
        disjunctive cause criterion, 94-97
        science of, 102
    definition, 11
    hotel booking cancellation rate, 66, 78, 84
context dependency of human behavior, 5
    contextual variables, 33
contiguity as contextual variable, 34
control versus treatment groups, 182
cookies for random assignment, 184
Cook's distance, 148, 162
correlation
    causation implied, 39, 78, 84
        causal mechanisms explained, 298
        coefficient of correlation as measure of
            causal effect, 8
        identifying moderators in historical data,
            272
        mediation, 298, 306
        white lies, 85
    forks creating appearance of, 50
    line of best fit, flat, 15
    missingness of missing data, 115-121, 129,
        130
    relationships between numeric variables, 78
coverage of confidence intervals, 192
Cramer's V for categorical variables, 81
CSAT (see customer satisfaction)
Cunningham, Scott, 63
curse of dimensionality, 7
customer behavior
    about instrumental variables, 297
    business behaviors affecting interpretation,
        27
    customer engagement, 31
customer experience (CX)
    cognition and emotions, 23, 30
    theory of change in booking profit versus,
        205-208
customer satisfaction (CSAT)

call center target metric, 234
    incentivization muddying, 235
cognition and emotions, 23, 30
measuring effect on future purchasing, 28,
    30, 31, 35
    business problem, 307
    data for example, 307
    instrumental variable measurement,
        311-314
negative impact from intervention, 205-207
CX (see customer experience)
cycles in causal diagrams, 57
    managing cycles, 58

**D**
data
    about causal-behavioral framework, 3
    about human behavioral data, xviii, 7
    about intuition, xii, xvii
    aggregate metrics, 32
    biases (see biases in data)
    categorical (see categorical variables)
    causal diagrams, 42-46
        building from scratch, 64
        error terms, 45
        probabilistic graphical models, 46
    connecting behaviors and data, 28
        behavioral variables refined, 32
        categorizing data, 30
        context understood, 33
        distrust and verify, 29
        existing data and legacy processes, 28
    extrapolation, 6
        human behavior data, 7
    GitHub for (see GitHub)
    hotel bookings real-world data set, 64
    influential points, 162
    interpolation, 6
    missing data (see missing data)
    modeling human behavior
        action or behavior data, 26
        business behaviors, 27
        cognition and emotions, 23
        demographic information, 22
        intentions, 24
    numeric (see numeric variables)
    outlier in small data, 148-154
    suboptimal, 147
        (see also Bootstrap; uncertainty)

variable selection examples, 8
  adding variables, 11
  adding wrong variables, 14
  causal diagrams, 67
  confounder in action, 9
  data for, 9
  multicollinearity, 13
  variables validated via data, 77-85
decision points in causal diagram building, 71
defiers, 229
demographic variables
  causal diagram building, 72
  data and ethical considerations, 22
  forks, 51
  high effectiveness of treatment, 265
  personal characteristics in behavior, 22
  primary causes, 21, 108
  social factors stronger than, 22
  stratification in stratified randomization, 211
  summary variable linear algebra proof, 56
dependent variable
  binary and logistic regression, 43
  causal analytics, 8
  mean via regression, 149
descriptive analytics, 4
Designing for Behavior Change: Applying Psychology and Behavioral Economics (Wendel), xiii
dimensionality curse, 7
disjunctive cause criterion (DCC)
  deconfounding first decision rule, 94-97
  definition, 94
  limitation of, 97
  sufficient but not necessary, 95
Downey, Allen, 186
dummyVars()
  one-hot encoding, 213
  package, 209, 236
duration
  contextual variables, 34
  time study, 148-154

**E**

econometrics text on instrumental variables, 308
effect sizes
  B.E.A.N. (beta, effect size, alpha, sample size), 188
  calculating, 277
  true/false positives and negatives, 186
effects nonlinear (see nonlinearity in effects)
Efron, Bradley, xi
emotions in human behavior
  causal diagram building, 71
  data and ethical considerations, 23
  measuring effect on behavior, 31
  model of behavior, 23
encouragement design, 224
  complier average causal effect, 226-231
  intention-to-treat (ITT) estimate, 224
engagement as behavior, 31
environment and behavior
  complexity of human behavior, 5
  feedback loop, 58
  social factors stronger than demographic, 22
ethics
  data collection for behaviors
    business behaviors, 27
    cognition and emotions influenced, 24
    intention influencing, 25
    misattributing effects, 22
    modifying behavior, 26
  New York Times test of intentions, 24, 26
  sludges, 24
exogenous variables, 108
experimental design
  about booking profit versus CX, 203
  about call center problem-solving mode, 233
  about four values of, 188
  about 1-click booking site, 173
  about terms used, 174
  bypassing political problems, 204
  customer satisfaction negative impact, 205-207
  data and packages for examples, 181, 208, 236
  encouragement design, 224
    complier average causal effect, 226-231
    intention-to-treat (ITT) estimate, 224
  hierarchical linear modeling
    about, 235, 237-242
    about business problem, 233
    analyzing the experiment, 253
    clustering variable, 238
    data and packages for example, 236
    planning the experiment, 234-236

power analysis, 245-252
power curve, 249
random assignment, 242-243
how many experiments, 188
planning the experiment, 174-181, 205-208, 234-236
behavioral logic, 179-181, 208, 236
best case scenario worth, 180
business goal, 175, 205-207, 234
importance of, 204
intervention, 177, 207, 234
target metric, 175-177, 205-207, 234
theory of change, 174
random assignment, 182-185
about business problem, 173
analyzing experimental results, 199-202
behavioral level of, 184
code implementation, 183
control versus treatment, 182
cookies for, 184
data and packages for example, 181
planning the experiment, 174-181
power/significance trade-off, 222
timing of, 182, 183
track for linking to outcomes, 184
sample size, 185-199
about, 182, 185
about statistics behind, 185
behavioral levels, 184
Bootstrap simulations for power analysis, 192-199
statistical power, 187-192
Test of Proportions for, 189
traditional power analysis, 189
stratified randomization
about, 203, 215
about business problem, 203
analyzing experimental results, 223-231
complier average causal effect for mandatory intervention, 226-231
data and packages for example, 208
intention-to-treat (ITT) estimate, 224
optimal algorithm, 236
planning the experiment, 205-208
power analysis with Bootstrap, 216-223
power/significance trade-off, 222
random assignment, 209-216
explain-away effect, 16
extrapolation, 6

curse of dimensionality, 7
Eyal, Nir, 26

**F**

factorials, 246
false positive probability, 186
mediation reducing, 297
moderation increasing, 270, 274, 275, 279
self-moderation limiting, 279
statistical significance, 186
target metric selection, 176
feedback loops, 58
fictional companies (see AirCnC; C-Mart)
forks in causal diagrams, 50-52
M-pattern, 101
noncausal paths, 99
paths, 61
slicing or disaggregating variables, 54
unknown forks with two-headed arrows, 51
frequency as contextual variable, 33
functional-style programming, xvi

**G**

gender as behavioral determinant, 22
generalized linear models (GLM), 42
Géron, Aurélien, xiii
GitHub
about standard packages, xiv
book code examples, xvii
causal diagram–building data, 64
CSAT and future purchasing, 307
experimental design data
booking profit vs CX, 208
call center as problem solvers, 236
1-click button site, 181
hotel bookings real-world data set, 64
missing data simulated data, 109
moderation analysis data, 258
variable selection data, 9
goals (see business goals)
granularity
actions and behaviors, 25
units of time, 81
Grolemund, Garrett, xiii

**H**

Hands-On Machine Learning with Scikit-Learn, Keras, and TensorFlow (Géron), xiii

hierarchical linear modeling (HLM)
  about, 235, 237-242
    clustering variable, 238
    Python code, 240
    R code, 238-240
  about business problem, 233
  analyzing the experiment, 253
  data and packages for example, 236
  package for HLM, 236
  planning the experiment, 234-236
    behavioral logic, 236
    business goal and target metric, 234
    intervention, 234
  power analysis, 245-252
    clustered data, 245-249
    permutations for limited randomness,
      246-249
    power curve, 249
  random assignment, 242-243
histograms
  Bootstrap regression analysis, 159
  Bootstrap validating moderation, 287
  count of samples of proportions, 156
  distribution of means of Bootstrap samples,
    152, 154
  personality traits, 135
Hooked: How to Build Habit-Forming Products (Eyal), 26
human behavior
  about human complexity, 5
  basic model
    about behavioral mindset, 20
    about five components, 21
    action, 25
    business behaviors, 26
    cognition and emotions, 23
    intentions, 24
    midlife crisis example, 20
    personal characteristics, 21
    user experience versus behavioral sci-
      ence, 23
  connecting data and behavior, 28
    behavioral variables refined, 32
    categorizing data, 30
    context understood, 33
    distrust and verify, 29
    existing data and legacy processes, 28
  data extrapolation, 7
  predicting future from past, 6, 7

hypergeometric distribution, 211

I

identity function I(), 269
influential points, 162
information unknown
  causal diagrams eliciting, 67
  contextual variables, 34
  unknown forks, 51
instrumental variables (IV)
  about, 297, 307
  applying, 314
  causal diagram roots, 308-311
  measurement, 311-314
  packages, 308
  understanding and applying, 308
intention-to-treat (ITT) estimate, 224
  complier average causal effect, 229
intentions
  about, 24, 70
  atomic versus aggregate data, 32
  causal diagram building, 70
  data and ethical considerations, 24
  intention-action gap, 24
  modeling human behavior, 24
    intent modeling, 25
  pain points as obstacles to, 25
interactions
  example, 257
  interacting moderators, 283-286
  machine learning capturing, 266
  moderated moderation as, 285
  moderation analysis, 265, 269
interpolation, 6
interventions
  booking profit versus CX, 207
  call center problem-solving mode, 234
  1-click booking site, 177
  testing smallest possible, 178
  theory of change, 174
An Introduction to the Bootstrap (Efron and
  Tibshirani), xi
intuition
  about data, xii, xvii, 46
  about what drives behaviors, 41
  from causal diagrams, 46
    deconfounding, 102

# K

Kahneman, Daniel, 67
Kaltenbrunner, Andreas, 211

# L

latent variables (see unobserved variables)
libraries (see packages)
linear algebra in causal diagrams, 42, 43
    aggregating variables and, 56
    transitivity property in collapsing chains, 49
linear regression
    causal diagrams, 44, 45
    line of best fit, 6
        flat lacking correlation, 15
    nonlinear relationships via, 267
    numeric effects, 274
    residuals
        Bootstrap, 163
        QQ-plot, 163
    stratified randomization analysis, 223-231
logistic regressions
    causal diagrams, 43
    generalized linear model, 42, 43
    residuals, 163
    statistical significance, 128
LTV (lifetime value) as target metric, 175
Lucas, Robert, 7

# M

M-pattern, 101
Mackenzie, Dana, 11, 16, 101
MAR (missing at random), 124
    diagnosing, 128
MCAR (missing completely at random), 124
    diagnosing, 126
McKinney, Wes, xiii
md.pattern() visualization function, 109, 110
mediation
    about, 297, 301
    about instrumental variables, 297
    biases in data and analyses
        causal biases, 299
        mediation unaccounted for, 297
    business problem, 298
    causal mechanisms explained, 298
    data for example, 307
    identifying mediation, 301
    instrumental variables

    about, 297, 307
    applying, 314
    causal diagram roots, 308-311
    measurement, 311-314
    packages, 308
    understanding and applying, 308
    measuring mediation
        about, 302
        binary variable mediator, 306
        direct effect, 305
        mediated effect, 303
        total effect, 302
    mediators in causal diagrams, 48
        slicing or disaggregating variables, 54
    moderation complement, 298
mental states (see cognition; emotions)
metrics (see target metrics)
mice package in R, 110
MinMaxScaler, 213
    package, 209, 236
missing data
    about, 107
    about example business problem, 107
    adding tracking variable for when missing, 123
    amount of missing data, 113-115
        amount safe to drop, 113
    biases in data, 107, 137
        assuming worst reducing, 126
    causality interaction unknown, 126
    cause diagnosed with CDs, 121-124
        adding missingness variable, 123
    correlation of missingness, 115-121, 129, 130
    data and packages for example, 109
    decision tree for diagnosing, 135
    fixing missing data
        about, 136
        adding auxiliary variables, 143
        MI predictive mean matching, 140
        multiple imputation, 137-139
        normal imputation, 141
        robust imputation, 143
        scaling up number of imputed data sets, 145
    missingness logistic regression
        Python, 127
        R, 127
    Rubin's classification of causes, 124-132

binary nature of, 132
fixes aren't recipes per classification, 137
missing at random, 124
missing at random diagnosed, 128
missing completely at random, 124
missing completely at random diagnosed, 126
missing not at random, 125
missing not at random diagnosed, 130
missingness probabilistic–deterministic spectrum, 132-135
statistical significance, 128
visualization of missing data, 110-113
  Python, 112
  R, 110
wrong or false values, 123
MNAR (missing not at random), 125
diagnosing, 130
model of human behavior
about behavioral mindset, 20
five components, 21
  action, 25
  business behaviors, 26
  cognition and emotions, 23
  intentions, 24
  personal characteristics, 21
midlife crisis example, 20
user experience versus behavioral science, 23
moderation analysis
about questions addressed, 257, 269
applying
  average effect redistributed, 276
  data analysis stage, 274-279, 298
  effect size calculations, 277
  experimental design stage, 270-273, 298
  false positives risk increased, 270, 274, 275, 279
  self-moderation alerting to hidden moderator, 280
  self-moderation false positives risk limited, 279
  standard deviation comparison, 274
  upper bounds on moderated effects, 276
  when to look for moderation, 270-286
data and packages for example, 258
definition, 258
interactions, 265, 269
  example, 257

machine learning capturing, 266
interpreting individual coefficients, 289-295
  calculating effects at decision level, 293
  setting meaningful reference points, 291
mediation a complement to, 298
multiple moderators
  interacting moderators, 283-286
  moderated moderation, 283-286
  moderated moderation as interaction, 285
  moderated moderation as segmentation, 285
  parallel moderators, 281-283
nonlinearity in effects
  about, 266, 269
  example, 257
  false positives risk limited, 279
  linear regression representing, 267
  moderated moderation applying to, 286
  self-moderation of nonlinearity, 268
segmentation
  about, 259, 269
  example, 257
  personalization, 264
  segmenting experimental data, 264
  segmenting observational data, 259
  uplift analysis, 264
validating with Bootstrap, 287-289
modifying behavior reference books, 26
Mostly Harmless Econometrics (Angrist and Pischke), 308
multicollinearity in adding variables, 13
multiple imputation (MI)
adding auxiliary variables, 143
introduction, 137-139
normal imputation, 141
predictive mean matchings, 141
Python code, 138
  categorical variables, 139
  predictive mean matching only, 140
R code, 137
  default imputation methods, 140
  ImputeRobust, 143
  normal imputation, 141
robust imputation, 143
scaling up number of imputed data sets, 145

## N

never-takers, 229

noncausal paths, 99
nonlinearity in effects
about, 266, 269
example, 257
linear regression representing, 267
moderated moderation applying to, 286
moderation self-positive effects limited, 279
self-moderation alerting to hidden moderator, 280
self-moderation of, 268
nonsharp null hypothesis, 186
Nudge: Improving Decisions About Health, Wealth, and Happiness (Thaler and Sunstein), xiii, 26
null hypotheses, sharp versus nonsharp, 186
numeric variables
binary variables with 0/1 values, 78
causal diagram variables, 77
mediator as, 306
missing data safe to drop, 113
relationships between, 78
relationships with categorical, 84
moderating the effect, 274
NumPy
Bootstrap optimization in Python, 169
importing as np, xiv
NYT (New York Times) test of intentions, 24, 26

**O**
1-click booking, 173
(see also random assignment)
one-hot encoding, 213
dummyVars() function, 213
package, 209, 236
OneHotEncoder, 213
package, 236
OneHotEncoder
one-hot encoding, 213
package, 209, 236
optimal algorithm for stratified randomization package, 236
outliers
hierarchical linear modeling handling, 237
regression analysis, 157
small sample size, 148-154
Overall Evaluation Criterion (OEC), 177, 205-207

**P**
p-value, 128
about, 160
beta (β) as, 188
Bootstrap to simulate, 159
false positive rate of zero impact, 186
5% common convention, 187, 188
p-hacking, 153
power/significance trade-off, 222
packages
about standard packages, xiv
as.binary() function, 236
block() function, 209, 236
Cook's distance, 148, 163
dummyVars() one-hot encoding, 209, 236
hierarchical linear modeling, 236
instrumental variables, 308
logistic() function, 109
md.pattern() visualization function, 109, 110
melt() function, 109
MinMaxScaler, 209, 236
multiple imputation, 109, 138
OneHotEncoder, 209, 236
optimal algorithm for stratified randomization, 236
power analysis, 181
QQ-plot, 148
random for sample() and shuffle(), 209
rescale() function, 209, 236
standardized effect size, 181
pain points as obstacles to intent, 25
pandas imported as pd, xiv
parallel moderators, 281-283
parent in causal diagrams, 47
identifying further causes, 87
paths in causal diagrams, 60
backdoor criterion, 98-103
blocked paths, 99
causal paths, 99
noncausal paths, 99
unblocked paths, 99
Pearl, Judea, 11, 16, 101
personal characteristics in human behavior, 21
causal diagram building, 72
data and ethical considerations, 22
high effectiveness of treatment, 265
personality changes over lifetime, 21
primary causes, 21, 108

personalized messaging, 264
Pischke, Jörn-Steffen, 308
power analysis, 181, 187-192
    Bootstrap simulations, 192-199
        code, 193-199
        stratified randomization, 216-223
    hierarchical linear modeling, 245-252
    power/significance trade-off, 222
        (see also statistical power)
    traditional power analysis, 189
predictive analytics
    about, 4
    causal analytics versus, 4
        regression demonstrating, 8
    complexity of human behavior, 5, 7
    interpolation versus extrapolation, 6
predictive mean matching (PMM), 141
    Python PMM only imputation method, 140
    R pmm method, 140
primary causes
    demographics as, 21, 108
    exogenous variables as, 108
probabilistic graphical models, 46
Python
    about book version, xv
    about packages, xiv
    about R versus Python, xiv
    Bootstrap
        confidence intervals, xvi, 154, 155
        mean of simulated samples, 151
        optimizing, 169
        p-value simulation, 160
        power analysis, 193-199, 216-223
        regression analysis, 158
    confidence intervals, 149
        Bootstrap code, xvi, 154, 155
        drawing with replacement, 151
    Cook's distance, 162
    Cramer's V function, 81
    hierarchical linear modeling
        about syntax, 240
        analyzing the experiment, 253
        permutations, 248
        power analysis, 245
        stratified randomization, 243
    mediation instrumental variable measure-
        ment, 311-314
    missing data visualization, 112
    missingness logistic regression, 127

moderation analysis
    identity function I(), 269
    interpreting individual coefficients,
        290-295
    moderated moderation, 285
    quartiles for numeric causes, 274
    segmentation, 262
    self-moderation, 268
    setting meaningful reference points, 291
    standard deviation of effect, 275
multiple imputation, 138
    predictive mean matching only, 140
p-value simulation, 160
random assignment, 183, 210
regression analysis, 157
regression residuals, 163
stratified randomization
    hierarchical linear modeling, 243
    intention-to-treat estimate, 224
    rescaling and one-hot encoding, 213
Test of Proportions for sample size, 190
Python for Data Analysis (McKinney), xiii

Q
QQ-plot, 163
    package, 148
quantiles for confidence intervals, 154
quartiles for numeric causes, 274

R
R
    about book version, xv
    about packages, xiv
    about R versus Python, xiv
    Bootstrap
        confidence intervals, xvi, 154, 155
        mean of simulated samples, 151
        optimizing, 167
        p-value simulation, 160
        power analysis, 193-199, 216-223
        regression analysis, 158
    confidence intervals, 149
        Bootstrap code, xvi, 154, 155
        drawing with replacement, 151
    Cook's distance, 162
    hierarchical linear modeling
        about syntax, 238
        analyzing the experiment, 253
        permutations, 248

power analysis, 245
    stratified randomization, 242-243
mediation instrumental variable measurement, 311-314
mice package, 110
missing data visualization function, 109, 110
missingness logistic regression, 127
moderation analysis
    identity function I(), 269
    interpreting individual coefficients, 290-295
    moderated moderation, 285
    quartiles for numeric causes, 274
    segmentation, 262
    self-moderation, 268
    setting meaningful reference points, 291
    standard deviation of effect, 275
multiple imputation, 137
    default imputation methods, 140
p-value simulation, 160
random assignment, 183, 210
regression analysis, 157
regression residuals, 163
stratified randomization
    creating pairs, 214
    hierarchical linear modeling, 242-243
    intention-to-treat estimate, 224
    rescaling and one-hot encoding, 213
Test of Proportions for sample size, 189
R for Data Science (Grolemund and Wickham), xiii
random assignment, 182-185
    about business problem, 173
    age moderating 1-click booking, 270
    analyzing experimental results, 199-202
    behavioral level of, 184
    code implementation, 183
    control versus treatment, 182
    cookies for, 184
    data and packages for example, 181
    pitfalls of, 183, 203
        (see also stratified randomization)
    planning the experiment, 174-181
        behavioral logic, 179-181
        business goal and target metric, 175
        intervention, 177
        poor target metric pitfalls, 176
    power/significance trade-off, 222
    timing of, 182, 183

track for linking to outcomes, 184
random package for sample() and shuffle(), 209
random seed via set.seed(), xiv
randomized controlled trials as causal analytics tool, 5
reactance, 229
regression
    about book, xii, xiii
    about book output, xv
    binary dependent variable, 43
    Bootstrap for regression analysis, 157
    coefficient of correlation, 8
        correctly structured regression, 8
    confidence intervals via, 149
    confounders introducing bias, 11
    line of best fit, 6
        flat lacking correction, 15
    moderation definition, 258
    numeric effects, 274
    parallel multiple moderators, 281
    predictive versus causal analytics, 8
    relationships between numeric and categorical variables, 84
    residuals
        Bootstrap, 163
        QQ-plot, 163
    statistical significance, 128
    times~1 with only intercept, 149
rescale() function package, 209, 236
resources for information
    book supplemental material online, xvii
    Nudge: Improving Decisions book, xiii
    Python for Data Analysis book, xiii
    R for Data Science book, xiii
RStudio, xv
Rubin, Donald, 11, 124
Rubin's classification of missingness causes, 124-132

S
sample size
    about, 182, 185
    about statistics behind, 185
    Bootstrap, 150
    experimental design, 185-199
        behavioral levels, 184
        Bootstrap simulations for power analysis, 192-199
        statistical power, 187-192

Test of Proportions for sample size, 189
   traditional power analysis, 189
N as, 188
stratified randomization increasing statistical power, 203
sample() function, 217
   package, 209
schedules as contextual variables, 34
scientific number notation canceled, xiv
segmentation
   about, 259, 269
   example, 257
   moderated moderation as, 285
   personalization, 264
      uplift analysis, 264
   segmenting experimental data, 264
   segmenting observational data, 259
self-moderation of nonlinearity, 268
   false positives risk limited, 279
set.seed() for random seed, xiv
sharp null hypothesis, 186
   confidence intervals, 192
Shipley, Bill, 39
shuffle() function package, 209
sludges
   business behaviors and customer behavior
      interpretation, 27
   NYT test failure, 24
social factors stronger than demographic, 22
social schedules as contextual variables, 34
Spyder (Anaconda), xv
standard deviation comparison in moderation,
   274
standard error calculated via regression, 149
standardized effect size package, 181
statistical power, 187-192
   1–α as, 188
   power/significance trade-off, 222
   sample size and, 188
   stratified randomization increasing, 203
   traditional power analysis, 189
      (see also power analysis)
statistical significance (p-value), 128
   about, 160
   beta (β) as, 188
   Bootstrap to simulate, 159
   false positive rate of zero impact, 186
   5% common convention, 187, 188
   p-value hacking, 153

power/significance trade-off, 222
stratified randomization
   about, 203, 215
   about business problem, 203
   analyzing experimental results, 223-231
      complier average causal effect for
         encouragement design, 226-231
      encouragement design, 224
      intention-to-treat (ITT) estimate, 224
   hierarchical linear modeling, 242-243
   optimal algorithm
      package, 236
   planning the experiment, 205-208
      behavioral logic, 208
      business goal, 205-207
      intervention, 207
      target metric, 205-207
   power analysis with Bootstrap, 216-223
      power/significance trade-off, 222
   random assignment, 209-216
      about, 209
      one-hot encoding, 213
      stratified randomization, 211-216
      without stratification, 209-211
subjectivity in causal diagrams, 41
substitution effects, 57
   cycles in causal diagrams, 57
Sunstein, Cass, xiii, 26
surveys
   call center CSAT target metric, 234
   intention data collection, 25
systems thinking, 58

**T**

T-test of means, 223
target metrics
   booking profit versus CX, 205-207
   call center problem-solving mode, 234
   leading indicators for, 175
   1-click booking site, 175-177
   operational versus financial, 176
   Overall Evaluation Criterion, 177, 205-207
   poor target metrics, 176
   theory of change, 174
   weighted average of metrics, 177, 205-207
Test of Proportions for sample size, 189
Thaler, Richard, xiii, 26
theory of change (ToC), 174
   behavioral logic, 179-181, 208, 236

moderator plus, 271
booking profit versus CX, 205-208
call center problem-solving mode, 234-236
1-click booking site, 174-181
Think Bayes (Downey), 186
Tibshirani, R. J., xi
time intervals
    aggregate metric hazards, 32
    granularity, 81
time study (time-and-motion study)
    about, 148
    about uncertainty, 147
    cake time study, 148-154
    introduction to Bootstrap, 148-157
timestamps for contextual variables, 33
timing
    experimental group assignment, 182, 183
    managing cycles, 58
    time trends in building causal diagrams, 75
traits of personality in causal diagrams, 73
transactional data, 26
transformations of causal diagrams
    about, 53
    aggregating variables, 55
    cycles, 57
        managing, 58
    paths, 60
    slicing or disaggregating variables, 54
transitivity property in collapsing chains, 49
treatment versus control groups, 182
true negative probability, 186
true positive probability, 186

**U**

unblocked paths, 99
uncertainty
    about, 147
    Bootstrap
        about business problem, 148
        introduction to, 148
    missing data introducing, 137
    moderation second-order effect, 287
unobserved variables, 122
    ovals in causal diagrams, 121
    proxies in causal diagrams, 86
    shaded rectangles or ovals in causal dia-
        grams, 40
uplift analysis, 264
user experience (UX)

behavioral science versus, 23
cognition and emotions, 23
intention observations, 25
UX (see user experience)

**V**

variables
    backdoor criterion, 98
    behavioral variable characteristics, 32
    causal diagrams, 40, 42
        forks, 50-52
        parent/child relationships, 47
        proxies for unobserved variables, 86
        transformations of, 53-61
        variables to include, 67-76
        variables validated via data, 77-85
    coefficient of correlation, 8
    confounder definition, 11
    demographic (see demographic variables)
    disjunctive cause criterion, 94
    existing data and legacy processes, 28
        categorize to understand, 30
        distrust and verify, 29
    exogenous variables, 108
    instrumental variables described, 297, 307
    mediation definition, 297
    missing data (see missing data)
    one-hot encoding categorical, 213
    predictive power with human behavior, 5, 7
    regression line of best fit predicting values, 6
    stratification in stratified randomization,
        211
    unobserved variables, 122
        ovals in causal diagrams, 121
        proxies in causal diagrams, 86
        shaded rectangles or ovals in causal dia-
            grams, 40
    variable selection examples, 8
        adding variables, 11
        adding wrong variables, 14
        causal diagrams, 67
        confounder in action, 9
        data for, 9
        importance of variable selection, 91
        multicollinearity, 13
visualizing missing data, 110-113

**W**

weighted average of metrics, 177, 205-207

Wendel, Stephen
    Designing for Behavior Change book, xiii
    granularity of actions, 25

Wickham, Hadley, xiii
working examples (see AirCnC; C-Mart)
Wright, Philip and Sewall, 308

## About the Author

**Florent Buisson** is a behavioral economist with more than 10 years of experience in business, analytics, and behavioral science. He currently leads the experimentation practice of the online marketplace Cars.com.

He previously worked for a French strategy consulting firm where he used economics and data analysis to answer complex measurement questions, for example building an index measuring the stability of agricultural policies in developing countries on behalf of the UN Food and Agriculture Organization. He also started and led for four years the behavioral science team at Allstate Insurance Company.

Florent has published academic articles in journals such as the peer-reviewed *Journal of Real Estate Research*. He holds a master's degree in econometrics as well as a PhD in behavioral economics from the Sorbonne University in Paris.

## Colophon

The animal on the cover of *Behavioral Data Analysis with R and Python* is the South American rattlesnake (*Crotalus durissus*). This species of highly venomous pit viper inhabits areas across South America, except for the high Andes and the far south. It also can be found in a few Caribbean islands.

These rattlesnakes are variable in appearance, generally with a pale underside and darker-brown diamond shapes or stripes on its back showing against the paler background. They feed on rodents as well as lizards. Adults can grow up to 6 feet long, and in captivity have lived up to 20 years. They breed seasonally, and females give birth to as many as 14 live young at a time.

It's estimated that around 400 people a year die from snakebite in the Americas, and the bite of the South American rattlesnake is known to be particularly lethal. Its venom has four main toxins: crotoxin, convulxin, gyroxin, and crotamine, which are used by the snake to capture and digest its prey.

Rattlesnakes often use their cryptic camouflage as a first defense and remain still at the approach of a larger animal; this strategy can result in humans being bitten because they've come too close or even stepped on the snake. Another defense is the source of their common name: the unique warning feature of the "rattles" on their tails. These are made up of keratin scales with multiple, loose layers, and when the snake uses a set of unique tail muscles to vibrate its tail, the dry layers hit each other and produce the well-known sound. Each time the snake sheds its skin, a set of rattles is added, making the number of segments one potential indicator (along with the snake's size and length) of the snake's age.

# O'REILLY®

## Learn from experts.
## Become one yourself.

Books | Live online courses
Instant answers | Virtual events
Videos | Interactive learning

Get started at oreilly.com.

Milton Keynes UK
Ingram Content Group UK Ltd.
UKHW051635070924
447942UK00006B/97